Your Complete Forecast 2019 Horoscope

Your Complete Forecast 2019 Horoscope

BEJAN DARUWALLA
With
NASTUR DARUWALLA

First published in India by
HarperCollins *Publishers* in 2018
A-75, Sector 57, Noida, Uttar Pradesh 201301, India
www.harpercollins.co.in

2 4 6 8 10 9 7 5 3 1

Copyright © Bejan Daruwalla 2018

P-ISBN: 978-93-5302-337-9
E-ISBN: 978-93-5302-338-6

The views and opinions expressed in this book are the author's own and the facts are as reported by him, and the publishers are not in any way liable for the same.

Bejan Daruwalla asserts the moral right
to be identified as the author of this work.

All rights reserved. No part of this publication may be reproduced, stored in a retrieval system, or transmitted, in any form or by any means, electronic, mechanical, photocopying, recording or otherwise, without the prior permission of the publishers.

Typeset in 10.5/13.7 Sabon at
Manipal Digital Systems, Manipal

Printed and bound at
MicroPrints (India), New Delhi

Contents

Aries (21 March–19 April)	1
Taurus (20 April–20 May)	20
Gemini (21 May–20 June)	39
Cancer (21 June–22 July)	58
Leo (23 July–22 August)	77
Virgo (23 August–22 September)	95
Libra (23 September–22 October)	113
Scorpio (23 October–22 November)	131
Sagittarius (23 November–22 December)	150
Capricorn (23 December–22 January)	169
Aquarius (23 January–22 February)	188
Pisces (23 February–20 March)	207

The Message of the Zodiac	227
The 'Mostest' of Everything	228
Gems for the Zodiac Signs	232
Personalities	234
World Horoscope 2019	240
The New Age Astrology	242
Blessings and Special Wishes	244
A Brush with Death	252
About the Authors	255
Important Announcement	257

ARIES

21 March–19 April

Aries is the sign of the Pioneer, the Explorer. You Arians are spontaneous, enterprising, pioneering, impetuous and impatient, sexy, turbulent, active, sometimes even violent.

Ganesha says one picture is equal to a thousand words. Therefore, you Arians must imagine that you are the original Zulu warrior – 10 feet tall – sinewy and strong. You are standing in a circle. You are hurling a big sphere right in the eye of the Sun. That is how daring and powerful you really are.

This means you are ambitious, have an enormous drive and the ability to forge ahead of all the others in business or profession. Your planet is Mars. Mars is the commandant of the army. Yes, sometimes you are hasty, eager, impetuous, wild, and therefore not logical. But each sign is made differently. Therefore, you can only be *you*, so to say. You are a fiery sign and the hands of the fire reach out to the sky. Nothing in life is final as you know, but your speciality is in physical exertion, affairs demanding audacity and courage, over structural iron or steelworkers, carpenters, timbermen, machinists, pharmacists, barbers, police, soldiers, hardware merchants. Mars governs the external sex organs, bladder, muscular system, head, face, sense of taste. Its metals are iron and steel; its colours, dark red and magenta; its flower, narcissus; its day, Tuesday. Shakespeare the greatest dramatist ever was an Arian but a shameless borrower, and I have the integrity to at least inform you about my sources.

This time I am following it up with your digital plan for the year so that you can focus very well on it in advance and thus prepare yourself by making a blueprint of the entire year. The digital plan for the year will be short and to the point, and though it is a general reading, it should apply quite a bit to you.

BLUEPRINT FOR THE ENTIRE YEAR

January: Parents, in-laws, work, rewards, family, the effort you put in.

February: Contacts and group activities, gains and joys.

March: Expenses, health issues, but God's grace and success in ventures, fine connections and collaborations.

April: Confidence, success, charm.

May: Finances, family, food, fortune, the four Fs.

June: Communications and contacts at all levels.

July: House, home, family.

August: Romance, children, creativity.

September: Work, health improvement measures.

October: Marriage, relationships, contacts, trips, ties and opposition.

November: Loans, funds, health, taxes, accidents, legal matters.

December: Publicity, publishing, fame, religious rites, matters to do with parents and in-laws – in short, being in the public eye.

Aries stands for enthusiasm, power and adventure. Ganesha says this is certainly the old monthly round-up but specially for 2019 this monthly round-up becomes very necessary and precise and proper because of the planets Jupiter, Saturn and Pluto helping you in every possible way. Therefore, though it is the old monthly round-up, it still has a very special, new, different, accurate guideline specially for 2019. Ganesha says Bejan is now eighty-seven. It is his voice of experience which says it specially for the year 2019. Let me explain to you that astrology is mainly about the right person being at the right place and at the right time and therefore this blueprint of the year will be of special importance and meaning to you.

We shall repeat it once, but because Aries is the first sign of the zodiac we are informing you that the following relationship *will apply* to each sign.

> Mercury – Vision
> Venus – Touch
> Jupiter – Smell
> Saturn – Hearing
> Mars – Taste

Ganesha asks for a look at the decans. Ganesha says each of the sun signs encompasses 30 degrees of the zodiac. These can be further subdivided into three decans of 10 degrees each. Therefore, although all those born between 21 March and 19 April are classified under the Aries sign, the traits will vary. So let's see what the decans say about you: if you were born between 21 and 31 March (the first decan), your ruling planet is Mars. This is an explosive decan. Excess of energy leads to quarrels and clashes. Control your temper and tongue. Sudden illness may befall you if you are between forty-two and fifty-six. Your twenty-eighth year (the year of Mars in Indian astrology) should be outstanding.

If you were born between 1 and 10 April (the second decan), the planetary ruler is the Sun, making you ambitious yet idealistic and kind-hearted, with qualities of leadership, as well as constructive creativity.

If you were born between 11 and 19 April (the third decan), whose ruler is Jupiter, you like the good things of life and don't mind paying for them. You revere learning, are generous and intuitive and would do well in any position of responsibility and trust.

You Arians will be lucky in 2019. Jupiter will be in your ninth angle from 9 March 2018 to 2 December 2019. Jupiter favours music of the plants, herbs, space, black holes and so on. Your sign Aries, strictly speaking, shows the last frontier in research, TV, Internet, Facebook, WhatsApp, industry, cars, aeroplanes, mining, website, iron and steel, forging, blasting of rocks and mountains for minerals such as copper, tin, gold, iron and so on. I am emphasizing this in different ways, so that you get the message loud and clear that 2019 is for research, discovery, expanding the boundaries of your heart, mind, body, spirit. Astrology is all about timing. And the timing for you to evolve, to develop, to grow will be just right in 2019 and 2020. Let me be as precise as possible. The period could well be a period of tremendous expansion, research, growth, maturity and capability – with more potential power, journey, immigration and pilgrimage than you will experience in your entire life.

Now that you know how exciting and powerful the period is, you should get ready to take your best shot by letting your arrows of desire fly from your bow of burnished gold. Surprisingly enough, your range is from industry to the fine arts. In the fine arts I mention Bismillah Khan, the shehnai king, Ravi Shankar, the sitar maestro and Charlie Chaplin, one of the greatest comedians of all times, besides, of course, our own first-ever Field Marshal S.H.F.J. Manekshaw, who defeated Pakistan though not in fine arts but another different art—the art of warfare.

Ganesha says, in simple terms Saturn means responsibility and duty. For you Arians, Saturn will be in your tenth house from 21 December 2017 to 23 March 2020. Therefore, like it or lump it, you will have to carry huge responsibilities. Luckily, you have Herculean or powerful shoulders. Therefore, you will succeed in your plans, wishes and ambition.

But I have a word of advice for you. Do not force others into your beliefs and ideals and dogmas. You are a born leader and Saturn certainly helps you in methods, organization, execution, management, law, order and religion. You should not force your opinions down the throats of your colleagues, servants, collaborators, marriage partner or friends. If you strike a balance between command and persuasion, power and flexibility, I guarantee you in the name of Ganesha that 2019 will be a strikingly good year for you. You will have power, perks, prestige, promotion and, most certainly, a mighty improvement in status and prestige. In fact, I see a big jump in all these matters. The only condition is keep your cool. You are welcome to command others. But please do not push them so hard that they lose their self-respect and dignity. Here comes 'why'.

Saturn in your tenth angle means power and push. But if you cross the limits, it certainly means a downfall, a big crash from a great height. I do not want you to do that. Luckily for you, in 2019 Saturn will be helped by Neptune, the planet of imagination and vision. Therefore, do think of yourself; but also think of others. Let me make it very clear that in any matter to do with organization, industry, handling large groups of people, social work, heavy complexes and plants, new projects and ventures, you will have the ability, the skill and the confidence to push ahead bravely and powerfully. This could be used as a tool or a wedge to gain your ends. In short, be venturesome, be brave but also respect the desires and wishes of others. Let me put it all in a different way. Be a great commander. But also be a fine listener. That is the secret. Rise is certain but fall remains in your own hands. Your ambition is great, your drive is tremendous. I agree to it. But try to control your ego and all will be well.

Your happiness quota will be 86 per cent.

SPECIAL BONUS

Ganesha says my devotee Bejan is introducing the topic of special bonus for all of you. This special bonus is placed upon Mercury. Mercury represents journeys, ties, trips, mobility, migrations and all types of communication. Venus represents love, beauty, the fine arts, polish and finesse, diplomacy and persuasion, joy, and a big hurrah to life itself. Therefore, I am combining both communication and joy for you and giving you the necessary dates.

For Aries, Venus signifies marriage, publicity and dealing with the general public. Therefore, Venus is a key planet. It will be favourable from 7 January to 3 February, 2 to 26 March, 21 April to 14 May (lucky and easy), 9 June to 3 July (a little bit of charm can do wonders), 26 July to 20 August (wish-fulfilment is possible), 5 September to 8 October (ties, trips and romance), 2 November to 25 November (at parties and functions you will be at your best), 20 December to 13 January 2020 (socialize and circulate).

For all creative activities and entertainment, hobbies, children, marriage, ties and legal matters, Mercury will be mighty important. Mercury will be favourable from 24 January to 9 February, 9 April to 6 May (particularly), 21 May to 4 June (good time to ask for favours and obligations), 27 June to 18 July (love and romance), 11 to 28 August (once again, love and romance), 14 September to 2 October (connections, ties, trips and a change in attitude as well as behaviour), 9 to 28 December (promotion and perks, journeys and research).

However, I must be true to my own self. This is only a solar-scopic reading based mainly upon the position of Sun. Therefore, it may not be as accurate as a personal horoscope. At the same time, I have put in a lot of effort, imagination and inspiration as well as intuition into it. In other words, I have tried my best. The results are up to our maker. Yes, sometimes miracles do happen. Therefore, I end on a positive note though I know my limitations very well.

KEY TO WEALTH BY NASTUR DARUWALLA

Arians are strong, blunt and ambitious and are not easily satisfied with the minimal. You must chant:

'*Shrim*'. Chanting this sacred word over 108 times would bring money in your bank account. Also, donate money to the poor.

WEEKLY FORECAST BY PHASES OF THE MOON

6 January: New Moon in Capricorn

Ganesha says, time to reach out to people and places by all means of publicity, transport and journey. It is also a good time to burnish your image and that of the company you represent. On the strictly personal level, marriage, house moves, ceremonies are confidently predicted.

14 January: Moon's First Quarter in Aries

You will work at full steam this period. However, Mars opposes your Sun sign and that could lead to a clash of will and competition. Take all the help from all the sources that you have and make a dash. It is important for you to take periodic rest, if you want to maintain good health.

21 January: Full Moon in Leo

While work pressures will still continue to mount, you will be entering into a phase which will have much to do with all sorts of partnerships and links. A journey may be essential to clarify policies, attitudes, situations and circumstances. These four will be the mainstay of the period. Try to make things simple and uncomplicated.

27 January: Moon's Last Quarter in Scorpio

Food, finance and family will be the triumvirate for you. Your stars will help you in terms of work, loans and settlements. It would be best to take a long-term view of men and matters.

4 February: New Moon in Aquarius

The focus shifts to love, children, hobbies, sports, and speculation. You could well sing *'My love, is like a red, red rose'* with the Scottish poet Robert Burns. This is a good time to play up all your strong points, be it your physical attributes and/or mental brilliance and/or your large-heartedness. It is also a time to step out in style. Explore avenues of amusement and entertainment.

12 February: Moon's First Quarter in Taurus

This period will give you energy and passion. The main trend will be of joint finance, loans, buying, selling, investing. Yoga, tantra, mantra and meditation could interest and even excite you.

19 February: Full Moon in Scorpio

In the fullness of your heart you will form, or rather start forming, beautiful friendships and connections since Jupiter, the harbinger of joy and good luck, has started rolling in your Sun sign. Group activities, socializing, camaraderie and children help you to be a happy human being.

26 February: Moon's Last Quarter in Sagittarius

This period will benefit you materially. Those in business and professions will come out right on top. It would be an excellent idea to hone your communicative skills, considerable though they are. You will see more of your neighbours and relatives; alternatively, you may be moving house/office, comments Ganesha.

6 March: New Moon in Pisces

The employment and new projects angle will be strongly slated. True, you will work with a will. True also, that you will succeed, but it will help you if you take good care of your health throughout the month of March. Pets, servants, subordinates and colleagues will figure largely in your scheme of things.

14 March: Moon's First Quarter in Gemini

Here is your chance to be a pundit or a research worker or a professor, laughs Ganesha. Journeys, ceremony and publicity are fated. It is also the perfect time to prepare for a campaign or a massive onslaught. A situation dealing with religion, moral values, law and order will take up much of your time.

March 21: Full Moon in Libra

This phase of the Moon juxtaposes nicely with both Mars and Saturn, helping you in doing some soul-searching. Now is the time when you can heal not only others but also yourself. Money will win its way into your pockets and rapidly fly away. Cuts both ways. A visit to clinics and/or social centres is not ruled out.

28 March: Moon's Last Quarter in Capricorn

This quarter of the Moon is well positioned with Venus, and therefore, renovation, decoration or alteration in the home or office/shop is likely. Buying and selling of property and vehicles also come under the bracket in this quarter. What you do now will have a cumulative effect in July.

5 April: New Moon in Aries

This phase of the Moon helps you to get organized. The reason is that only by being organized and working to a tight schedule will you be able to complete a multitude of activities. This is the right time to approach people, to get your work done. Marriage and/or alliances and/or collaborations are not ruled out.

12 April: Moon's First Quarter in Cancer

This quarter is ideally placed with Saturn and it will help you to manage things well and thus, very obviously, get more mileage. It is a good time to lay down policies and procedures. Methods and systems can be changed and improvised. Yes, you will be working hard (and so am I, with these forecasts!)

18 April: Full Moon in Libra

Ganesha says this phase of the Moon makes you confident and even aggressive. It is an ideal time to launch a project, get engaged or even married, take a journey, improve your computer skills, enjoy sports, call friends over. Many of you will be striking out on your own

26 April: Moon's Last Quarter in Aquarius

As this quarter of the Moon is in fine combination with Mercury you will have the finance, the support and the wherewithal to expand or start a project of your choice. Children, hobbies, entertainment and amusement keep you busy as well as happy. Great going.

4 May: New Moon in Taurus

The excellent position of stars helps you in terms of romance, affairs of the heart, trusts and funds, joint finance. It is a fine fusion of pleasure and profit. If you are appearing for a test or an examination, you will score well. Religious rites are foretold.

12 May: Moon's First Quarter in Leo

You will be improvising in whatever you do. Inventors, mechanics, fashion or rocket and missile designers, potters and poets, scientists and surgeons will be in their element, a cliché but apt. Your efforts will be amply rewarded.

18 May: Full Moon in Scorpio

The Moon highlights your possessions, for example, cars, clothes, jewellery, shares, house; and if you are the jealous type, your better or bitter half, as the case may be. Ganesha says, make haste slowly because too much will be at stake. Home conditions will keep you on your toes.

26 May: Moon's Last Quarter in Pisces

Matters to do with servants, colleagues and even the big boss should turn out well for you. Therefore, expect promotion and perks. Pets can be acquired, so also can loans and funds. These frantic activities could take a toll on your health. Periodic rest is therefore essential.

3 June: New Moon in Gemini

You will break the shackles which have bound you and be like an eagle soaring in the sky at your leisure and pleasure. A free spirit. A journey, ceremony, publicity, meets, conferences, interviews – yes, all these are meant just for you. Plenty of movement in every imaginable direction!

10 June: Moon's First Quarter in Virgo

After that amazing spell of freedom during the last period, you will feel a little tired. You will turn the searchlight within yourself. Even in a crowd, you will feel lonely, though not lost. This is a good time to begin any creative activity, right from cookery to writing a novel to climbing mountains.

17 June: Full Moon in Sagittarius

The placing of the stars helps you in two different ways: a) to tide over difficulties and obstacles, and b) to take the lead in matters related to group activity, coordination, conferences, socializing and so on. This is a good time to set the pace. You can be both a leader and a follower, paradoxically.

25 June: Moon's Last Quarter in Aries

You will be working systematically and magnificently. The result is rich rewards and tremendous inner satisfaction. Because you will be appreciated, you will in turn appreciate others and spread good cheer. Doctors, nurses, social workers, psychiatrists, healers will have an extra spirit to their work.

2 July: New Moon in Cancer

You will kick off in great style in your work. The surprising part is that you will show the same enthusiasm in your play. Therefore, you will be leading a full and rich life. There will be so much on your plate that it will be difficult to rest and relax. But do that you must.

9 July: Moon's First Quarter in Libra

Jupiter has just changed signs, by Western astrology, and it will help you to come to terms with your surroundings and the domestic scene. Let me tell you quite bluntly that this will be the main trend for the period. Your income will shoot up. Personal matters will also clamour for attention.

18 July: Full Moon in Capricorn

The main trend of house, home renovation, family, possessions, interaction with parents and in-laws will accelerate to supersonic speed. So much will happen so fast that even thirty-six hours in a day may not seem enough – no exaggeration intended. Buying, selling, renting and leasing will also have to be taken care of in the next period.

25 July: Moon's Last Quarter in Taurus

Partnerships and ties and evidently new faces as well as relationships should make for exciting fare. You, luckily, thrive on excitement, so it is okay. You should cultivate assiduously the public relations and diplomatic side to your nature. Ganesha says that is absolutely vital. The home and family angle will also have its full impact.

1 August: New Moon in Leo

The Moon conjuncts Uranus and Neptune bringing out the wayward, rebellious but exceptionally energetic and creative side of your nature. All the world loves a lover, except the husband/wife, and you will be the great Laila/Majnu of romance. Those in electronics, photography, computers, the latest gizmos, show their extraordinary skills and talents. In short, time to be inventive and different.

7 August: Moon's First Quarter in Scorpio

Socializing, group activities, being a perfect host and a good guest will win you plaudits. Money is important and will be there for you, so rejoice. Hard work, tough competition, final success and commensurate rewards – these are the way it will be in August, says Ganesha. It all starts now.

15 August: Full Moon in Aquarius

The phase of the Moon helps to take decisions, and right decisions lead to triumph. The period will encourage socializing, friendship, introspection, matters related to children, finance, rentals, taxes, entertainment and banquets. Also, it will give you confidence and conviction.

30 August: New Moon in Virgo

The focus now swivels to the house, home, property, buying, selling, renovating and decorating. Once again, rentals, taxes, leasing, hypothecating take top priority. Possibly the health of elders, parents, in-laws, causes you acute anxiety.

6 September: Moon's First Quarter in Sagittarius

This quarter of the Moon gives you a great insight into the use, value and actual handling of high finance. The trend of the last period gathers great momentum. A home away from home, issues of passion, and sex, will be a dominant trend. Lottery and legacy could make your day.

14 September: Full Moon in Pisces

While hard work and a change in the nature and scope of your occupation are predicted, it will literally pay to take good care of your health. Otherwise you may have to pay through your nose for medical bills. Learn to communicate and contact efficiently and instantly. That could well be the difference between success and failure. Short trips and visits and interviews keep you busy.

22 September: Moon's Last Quarter in Gemini

Moon encourages you to go flat out in your work or even do a job switch. You might be visiting clinics, hospitals and social centres. Your handling of chores or nit-picking with neighbours and relatives could get you into trouble. Health and diet need special care.

28 September: New Moon in Libra

Mercury and Neptune help you take the initiative by better and faster communication. Your hunches or guesses or psychic ability will guide you in the right direction. Get your travel kit ready. Be friendly with foreigners and visitors

5 October: Moon's First Quarter in Capricorn

The Moon makes you friendly and outgoing. Many of you will be signing important deeds and documents. Partnerships, right from the professional to the platonic, are confidently predicted. So also are engagements and collaborations.

13 October: Full Moon in Aries

While the main trend will be of contacts and ties in October, this specific period finds you taking an unusual interest in family and home. This is a good time to buy, sell and shop around for anything you really need. Speculators and investors could take calculated risks, if so inclined. You will love others and in turn be cherished by them.

21 October: Moon's Last Quarter in Cancer

Expect drama and high tension because the Moon will trigger and target it. It would be in your interest to reach out splendidly in all possibly directions – sales, publicity, advertising, telephone calls, posters, email, taxes and so on. Once again, trips and building bridges of understanding are predicted. A journey with a stopover is very much on the cards.

28 October: New Moon in Scorpio

Gird up your loins and give it your best shot. Whether a peon, a professor or a prime minister, you will be forced to go that extra mile and carry your office work right into your home. In all fairness, you can expect excellent rewards for it later on. This is particularly valid for all professionals. Parents, in-laws and relatives will also lay claim to your time and money.

4 November: Moon's First Quarter in Aquarius

The Moon conjuncts between Jupiter and Saturn. It will give you a fine balance between idealism and practicality. The main issue will be an alignment or coordination of finance with family and food.

12 November: Full Moon in Taurus

This period makes you romantically and artistically inclined. Therefore, this is a good time to start writing, painting, singing, dancing, sculpting, designing, modelling, photography, acting – if that had been your desire all along. Friends and well-wishers, and even patrons will support you.

18 November: Moon's Last Quarter in Leo

The Moon connects well with Mars, thus helping you in securing a loan, a raising capital or investing/buying/selling/leasing/borrowing. If you are interested in the occult, yoga, tantra and mantra, necromancy or playing the planchette, this is the time for it.

26 November: New Moon in Sagittarius

The Moon makes you tolerant of strangers and foreigners and thus you won't suffer from xenophobia, i.e., hatred of foreigners and strange people. Trips and ties, communication, computing and contacts from the basis of your life.

4 December: Moon's First Quarter in Pisces

The Moon fuses with Pluto, the powerhouse, giving you pep, vivacity and go. Ceremonies, publicity, alliances, higher learning, great enthusiasm, exceptional ability to lead people are the predictions for you, in the name of Ganesha. And this will form the trend for the period.

14 December: Full Moon in Gemini

The Moon's period helps you to meet people in general and be a crowd-puller. Relaxed pleasure will be possible despite tremendous pressure of work, a contradiction which will come to pass. Please view money spent as entertainment. A new friend comes into your life.

19 December: Moon's Last Quarter in Virgo

The Moon helps you in attending to the needs and value judgements of household members. If interested in bettering your surroundings or beautifying your home, do go right ahead. You might engage in fix-it projects. Take advantage of an opportunity to get what you want.

26 December: New Moon in Capricorn

Relationships is the name of the game. Surprisingly enough, you'll be both breaking and following the rules. Go that extra mile for your sweetheart or friend and see a smile light up their faces. Be ready to adapt and adjust in minor matters. And may the new year bring a handsome harvest for you. Ganesha nods.

Taurus

20 April–20 May

Taurus is the sign of the Builder, the Producer. You Taureans are kind-hearted, pragmatic, materialistic, practical, artistic, self-indulgent, careful, trustworthy, patient and stubborn.

Ganesha pats his big stomach, laughs loudly and comments that Taurus the bull has many meanings. The first is bull in a China shop which means the bull becomes angry and disturbs many people. Secondly, in the stock market there is a phrase called bullish. It means a rise in the share market. Finally, we have Nandi, the bull of Shiva, and Varasyaji, the sacred bull of the Parsees, for worship. Ganesha says the bull therefore has an extra significance to it. The Parsees and the Hindus wear the sacred thread. The Parsees put the urine of the bull Varasyaji in their mouth before wearing the sacred thread. That is the real respect and sacredness the Parsees give to the bull. This bull is kept by the high priest and is completely white.

This time I am following it up with your digital plan for the year so that you can focus very well on it in advance and thus prepare yourself by making a blueprint of the entire year. The digital plan for the year will be short and to the point, and though it is a general reading, should apply quite a bit to you.

BLUEPRINT FOR THE ENTIRE YEAR

January: You leapfrog to fame, publicity, spirituality, fulfilment, journey, education, future plans, relations.

February: Tough decisions, health of elders, parents, in-laws, work premises; absolutely tremendous issues of prestige and status are possible.

March: Friendship, the social whirl, romance, material gains, hopes, desires, ambition; happy days are here again.

April: Expenses, losses, contacts, love, secret deals, journeys, spirituality; therefore, April will be a paradox, a big contradiction.

May: Confidence, power, gains, happiness, right timings, the realization of wishes.

June: Finances, food, family, buying/selling/shopping, property, functions and meets; you will be a crowd-puller! That's great.

July: The three Cs: contacts, communication, computers.

August: Home, house, parents, in-laws, property, requirements for the elderly, foundations for new projects.

September: Entertainment, love, engagement, hobbies, sports, games of chance – in a word, creativity.

October: Health, employment, pets, subordinates, colleagues, debts and funds.

November: Marriage, legal issues, friends and enemies, trips and ties, collaborations, competition – it is a mixed bag.

December: Money, passion, joint finance, buying/selling/shopping, taxes, real estate, insurance, focusing on health, and strain and drain (on the purse).

This time we are giving you extra information about your own sign. It will be useful forever and ever. Ganesha asks for a look at the decans. Ganesha says each of the zodiac signs is of 30 degrees, which can be divided into three decans of 10 degrees each. Thus, though all born between 20 April and 20 May are Taureans, their personality traits will vary, according to the decans.

If you were born between 20 April and 1 May (the first decan), your planetary ruler is Venus. Your nature is pleasure-loving and sensual and you must learn to exercise self-control. You can blow hot and cold to manoeuvre others. You will shine in pursuing the performing arts, theatre and the cinema and also in painting, hotel business and catering.

If you were born between 2 and 11 May (the second decan), your ruling planet is Mercury and you have a wide range — art, law, teaching, editorship, diplomacy, chemical work, salesmanship — to do justice to your undeniable talents and versatility. Your inherent caution helps you to rush ahead but to cover your bets at the same time, giving you skill and finesse in your moves over others.

If you were born between 12 and 20 May (the third decan) your ruling planet is Saturn. There is an inflexibility and rigidity in your approach. It's impossible to dislodge an idea that has taken root in your mind. Although this makes for single-minded devotion, it also sees you in a rut. Property, land, agriculture and gardening (all connected with the earth), engineering, entrepreneurship and administration are some viable pursuits.

Taurus and Scorpio are recognized as moneymaking signs by Western astrology. In 2019 you will find out the accuracy of it. Why? Jupiter the money planet will be in your eighth angle. The eighth angle signifies that:

- Finances, money and legal matters, whether dealing with wills and legacies, real estate and codicils, insurance, joint finances, transits and inheritance will be of paramount importance and occupy centre stage in your activities and preoccupy you.

- It does not mean that you do not indulge in power plays, passionate encounters, a bit of speculation – in short, spills and thrills.

- Raising capital, buying and selling stocks and shares, or just making canny investments, shopping for long-term gains, will all thrill and interest you.

The eighth house also stands for trusts, shares, investments, public money, mutual funds, midcaps, insurance, fixed deposit, legacy, inheritance, tantra and mantra, and prayers for the dead. Now you will understand that the eighth house is significant for money, property, wills, and sometimes the death of partners. Jupiter will be in your eighth angle from 9 March 2018 to 2 December 2019.

Saturn the other main planet will be in your angle of *Bhagya* or good fortune from 21 December 2017 to 23 March 2020. It will help you in:

- Higher learning and evolution
- Journey and ceremony
- Publicity, communication, contacts, computer knowledge and skills
- Inspiration and intuition
- Choosing the right time to plan well ahead (and believe me, this is the master key to success)
- Dealing with your in-laws and relatives, (whether you like it or not), and inducing you to get on with it charmingly (which is your wont)
- Trying to be a creative listener too (though I know this is difficult for you)
- Avoiding neglect of contacts and correspondence (which may lead to fatal consequences) or getting into needless arguments about trifles, just to prove that you are right (leading you to remain detached for your own good)
- Being firm but not inflexible and rigid to a fault (which you should be as suggested earlier)

Lastly religion, research, spirituality, pilgrimage, pious deeds, tours and travels and ties, journeys and all matters to do with your higher self and your consciousness will be mighty important. In other words, it is time for a complete renovation and evolution of your own self. This is truly a mighty powerful and necessary asset for your own personality and development.

Your happiness quota will be 86 per cent.

SPECIAL BONUS

Ganesha says my devotee Bejan is introducing the topic of special bonus for all of you. This special bonus is placed upon Mercury. Mercury represents journeys, ties, trips, mobility, migrations and all types of communication. Venus represents love, beauty, the fine arts, polish and finesse, diplomacy and persuasion, joy and a big hurrah to life itself. Therefore, I am combining both communication and joy for you and giving you the necessary dates.

Venus is the pilot of your jet. Therefore, when Venus turns favourable you will be at your best. The periods are 4 February to 1 March (journeys and pilgrimages), 27 March to 20 April (socializing and fun), 15 May to 8 June (perhaps it is a time when like a big game hunter you too make a kill – be it films, music, architecture, profit, business and so on), 4 to 27 July (contacts and communication), 14 to 21 August (love, romance, hobbies, entertainment, creativity), 9 October to 1 November (hectic romance, love, possibly marriage), 26 November to 19 December (you will be able to sum up that life has been well worth living).

Finances and family come under the direct control of Mercury. Mercury turns favourable from 5 to 23 January, 10 February to 16 April, 7 to 20 May (very particularly), 6 to 26 June (good for contacts and contracts), 29 August to 13 September (travel, trade, ties) and once again 29 December to 6 January, 2020 (you will end the year with a knockout blow to your enemy).

However, I must be true to my own self. This is only a solars-copic reading based mainly upon the position of Sun. Therefore, it may not be as accurate as a personal horoscope. At the same time, I have put in a lot of effort, imagination and inspiration as well as intuition into it. In other words, I have tried my best. The results are up to our maker. Yes, sometimes miracles do happen. Therefore, I end on a positive note, though I know my limitations very well.

KEY TO WEALTH BY NASTUR DARUWALLA

Taureans are responsible people; they shoulder all kinds of responsibilities – familial and societal. For them the mantra is: '*Om Sarvabadhai vinirmukto dhana dhanya sutanvitah manushyo matprasadena bhavishyati na sanshayah*'. Donate food packets to the needy people.

WEEKLY FORECAST BY PHASES OF THE MOON

6 January: New Moon in Capricorn

This phase of the Moon could result in publicity ventures, journeys, the birth of children or of fresh new ideas, flashes of intuition. Religious rites and rituals and changes on the home front are also forecast. I see a quickening of pace, bonding which may be short-lived or permanent, but there is certainly plenty of movement. It may even spell out the end of a relationship followed by the launch of another. Interesting, even exciting.

14 January: Moon's First Quarter in Aries

You are contemplative and highly introspective. A human relationship and an overseas angle will claim you. Journeys, collaboration, future plans and ambitions are in the forefront, and liaisons and meetings, even a romantic rendezvous, are probable. Expenses will shoot up rapidly but this is a passing phase.

21 January: Full Moon in Leo

The phase of the Moon sees you making things happen. Contracts are bagged, deeds and documents are executed, tests will be taken successfully. It is transport and communication time, so also time for meeting people, solving puzzles, acquiring higher learning, putting your ideas across. With knowledge, may well come the benefits of wisdom and spirituality, the rewards of your past karma.

27 January: Moon's Last Quarter in Scorpio

This period gives you a certain closeness with people, in several ways. You excel at collaborations and ties, and at a one-to-one relationship with your soulmate. It may be a time to mend fences, to kiss and make up, since people and places motivate you. New assignments – with a substantial benefit – could fall into your lap.

4 February: New Moon in Aquarius

The phase of the Moon finds you making a huge effort to achieve, to assert, to score, to reach targets as dignity and respect mean the world to you. Matters related to prestige, status, work, reputation and honour will claim top priority. You will seek approval for future plans, discard wrong opinions. Distant relations and in-laws also clamour for attention; so if you do not acquire the gentle art of relaxation your health could even be undermined.

12 February: Moon's First Quarter in Taurus

This period emphasizes the personal and financial angle. Sex and salvation, birth, death and regeneration are strong indications; thus, children, creativity, news and views get a tremendous boost. You will come alive in every way, not least of which is your leanings towards yoga, meditation, tantra and mantra.

19 February: Full Moon in Scorpio

This phase of the Moon – and the trend for the period – turns the spotlight on to prestige, status, awards and rewards, anything to do with ambition, in other words. Equally important are home and family as extensions of yourself. Thus, the focus is equally on personal advancement, better prospects, making you busy with activities involving superiors and those that may make your ratings soar upwards on the popularity charts!

26 February: Moon's Last Quarter in Sagittarius

Your parental ties and duties, alongside your business and profession as well as likely changes in the house/home/office all give you a jump start. In short, the focus is on all your activities, and you may get honour and prestige and make progress. But remember, with achievement there can also be failure. It may seem like one step forward and one-and-a-half back, but stick with it to become a winner!

6 March: New Moon in Pisces

The Moon connects superbly with the three Fs of finances, funds and family. There is a great joie de vivre, a gusto and enthusiasm in all you do that sees you socializing, meeting some people, loving others. A strong likelihood of marriage or a liaison or a romantic attachment is an added bonus, I hope.

14 March: Moon's First Quarter in Gemini

You reach out to meet new and old friends and acquaintances and there is a chance of easy money as well as happiness in life. The two don't always go together, do they? Attachments and ties, children and sweethearts all clamour for your time, as do group activities, entertaining and being entertained, even seeing the realization of a long-cherished hope or dream.

21 March: Full Moon in Libra

This phase of the Moon enthuses and energizes you considerably. The artistic side of your nature will shine forth, so you will be a great lover, a fine parent, a scintillating actor/entertainer, the original idea person. Sociability and entertainment, sources of income other than honest-to-goodness hard work, and as said for the last period, all ties and attachments keep you in a distinctly upbeat mood.

28 March: Moon's Last Quarter in Capricorn

This quarter has you in a mood of nostalgia and sentimentality. It makes you feel a shade vulnerable, but also very humane and caring. A bit of meditation, along with house/home/family, rituals and religion, even travel and children all come together for you. Great going!

5 April: New Moon in Aries

The emphasis is on your angle of restlessness, soul-searching, spirituality. It is or can be reflected in journeys, children, responsibilities, accidents and perhaps bad health. Expenses have to be coped with. Do not neglect dear ones and/or offend well-wishers. Displaying a little restraint and altruism pays tremendous dividends.

12 April: Moon's First Quarter in Cancer

The ingress of Jupiter, the planet of beneficence, may be the turning point, so that things start looking up for you in terms of social activity, group involvement, forming close bonds with others, even with sweethearts and loved ones. Sudden gains or a promotion may also materialize – some form of easy money, in short. In this too there may be a tendency to go too far soon, so you'll have to watch it, warns Ganesha.

18 April: Full Moon in Libra

The Moon makes you work hard, play hard. While you have your full quota of duties and responsibilities, your efforts will get their due appreciation, even rewards. There may be some improvement in health as the stress factor is lower. Several Ps – projects, policies, promotion and perks, and pets – all clamour for your time and attention and manage to keep you busy.

26 April: Moon's Last Quarter in Aquarius

May is certainly your pivotal month. This period awakens an ultra-sensitivity, a yearning to be loved, cherished and appreciated. Alongside is a feeling, a restlessness, self-analysis and soul-searching. Loans and funds, insurance and joint finance are also important.

4 May: New Moon in Taurus

This phase of the Moon is the real time for trips, relationships and even unusual sex, different forms of stimulation. Finances, loans, coming together or parting permanently is the astrological message which is capable of several interpretations, depending on the individual placing.

12 May: Moon's First Quarter in Leo

This quarter of the Moon highlights the domestic scene – improvements in the home/office/workplace come within its ambit and so also do buying/selling/leasing. You will be dealing with confidential and private matters. Visits to lonely places, hospitals, even prisons could materialize too.

18 May: Full Moon in Scorpio

This is the time to receive, accept, interface with as many people as possible. There can be an increase in socializing, respectability, a sense of well-being varying with a feeling of self-undoing. Welfare interests and charities help you achieve a balance, but do try not to make the world's problems your own and don't give in too easily to these mood fluctuations.

26 May: Moon's Last Quarter in Pisces

There is a possibility, in this quarter of the Moon, of a surge in finances but with it, more responsibilities, especially to society. You may be into reiki or healing, deeply intuitive, even prone to hallucinations, and can experience low energy levels. Yet there may also be a collaboration, a permanent alliance, even marriage, or work-wise, a position of responsibility.

3 June: New Moon in Gemini

There is an improvement in terms of hope, comfort, bonding and money – practically all the important things of life. Much however, depends on you to make it come to pass; so a bout of soul-searching and self-analysis is indicated. Also behind-the-scenes and surreptitious activities, and a continued interest in welfare and charity. This phase of the Moon focuses more on future welfare.

10 June: Moon's First Quarter in Virgo

Your more imaginative and softer side comes to the fore. Hobbies can almost become a vocation as you are immensely creative and caring. A rare kind of sympathy may be experienced and you mesh beautifully in terms of bonding, romance and children.

17 June: Full Moon in Sagittarius

This quarter of the Moon brings you a charge of tremendous energy. Vitality will be at an all-time high, so that there are major changes in your outlook and environment. There are also financial benefits to be had. All this month, you may feel ready to tilt at the windmills and say, 'One day I'll change the world.'

25 June: Moon's Last Quarter in Aries

You will have to keep all channels of communication buzzing – whether personal, professional or otherwise. This quarter also favours relationships, liaisons, partnerships and rendezvous. You will be a natural leader and initiator, but take care to avoid going to emotional extreme.

2 July: New Moon in Cancer

This phase of the Moon concludes the trends of the month, and confidence oozes out of you and colours your personal outlook on the world. An exceptionally stimulating week lies ahead for you and there is a strong desire for action coupled with an equally strong desire to burnish your image in the eyes of others.

9 July: Moon's First Quarter in Libra

A fairly good time for food and finances. You will have to guard against loss, pilferage or being careless with important documents. Health care is necessary, particularly in terms of not taking on extra responsibilities and learning to relax and de-stress yourself.

18 July: Full Moon in Capricorn

Tests, interviews, ceremonies for the living and the dead will come to fruition. It will be a good idea to go all out for publicity ads, using all channels of communication. The spoken word and to a lesser degree, the written word, music and the visual arts are all focused; so also are family life and food; whether entertaining guests or going on a diet will remain to be seen!

25 July: Moon's Last Quarter in Taurus

What you do or have done during this period could well fructify in August. Long-range planning, looking ahead, have an extra significance and meaning. You will focus on finding ways to enchance your property and belongings, but guard against looking for quick profits or money from sources other than hard work.

1 August: New Moon in Leo

You will communicate remarkably well, in a continuation from this period. But if you get work done, you will also have your work cut out. Secret and sudden help cannot be ruled out. The range will be from marriage to collaborations, from a short journey to immigration. The stress will be on people coming into your life – and this is the monthly trend ushered in by the Moon.

7 August: Moon's First Quarter in Scorpio

A time to lay foundations for new projects, not least of which is cementing and strengthening the bonds of family and relationships. A change of environment may manifest itself in renovation of the house or office and some alterations which enhance or change your lifestyle. Some pending matters may be concluded, says Ganesha.

15 August: Full Moon in Aquarius

Expect visitors as meeting relatives, siblings, friends are all indicated. The theme will include communication at all levels. Also, in case of, older person's retirement may be on the cards and with benefits! You will forge ahead bravely, full of both creativity and the insights gained from experience.

30 August: New Moon in Virgo

The change of ambience theme will broaden to now include the world of spirituality and of a certain intensity in your perception of life. You will perceive a change in your psychological make-up, but do avoid stirring up trouble or making changes that may prove to be unsettling in the long run.

6 September: Moon's First Quarter in Sagittarius

The Moon blesses you with many and various things – foresight, artistic excellence, joy from children and a sense of fun. You will unravel complicated problems and situations. A great composure and cheerfulness, light-heartedness of spirit will not only make you happy, but enable you to spread happiness as well. Ganesha chuckles!

14 September: Full Moon in Pisces

This quarter helps in funds, finances and loans. Tantra and mantra and prayer may well yield you both pleasure and profit. On the one hand, there is entertainment and fun, and on the other, risk taking and high adventure in the world of finance. It's as thrilling as a Formula One race, I assure you.

22 September: Moon's Last Quarter in Gemini

The Moon hits the high spots, setting you firmly on the way to progress, achievement, recognition. The monthly trend initiated favours both work and enjoyment. Marriage and work alliances too, so that contrary influences and concerns do come to a meeting point. Publicity meets, conferences, seminars may well find you as the star performer.

28 September: New Moon in Libra

This quarter helps you in finances, loans, tantra and mantra, occult phenomena or experiences, prayer and spirituality, the rewards of your past karma. The material to the spiritual plane come within its spectrum. Buying/selling/shopping also come under its influence.

5 October: Moon's First Quarter in Capricorn

This quarter of the Moon spurs you into terrific activity. You experience both freedom and pleasure and there is both romance and joy with loved ones and children on the one hand and entertainment on the other. Creative expression will be your forte even as you strive for greater personal liberty. Guard against harbouring wrong notions or hurting the feelings of those you love and cherish.

13 October: Full Moon in Aries

The trio of family, finance and travel is what this quarter brings you. You catch up on both information and knowledge. There is a difference. And with knowledge comes wisdom which is reflected in your actions as well as attitude to life.

21 October: Moon's Last Quarter in Cancer

This quarter will revitalize and energize you afresh. That is the promise Ganesha holds out. Your family life benefits greatly, especially where parent–child relationships are involved. Justice and fair play a sense of duty are all important issues this week. You will play a vital role in resolving a conflict and there may almost definitely be a family get-together.

28 October: New Moon in Scorpio

Expenses, some travel and, almost certainly, behind-the-scene activities are what this quarter has in store for you. The other strong theme with this quarter right through the month will be dependents – whether colleagues, subordinates, servants or even pets. Do not let others' problems overburden you or you may fall victim to a kind of discontent.

4 November: Moon's First Quarter in Aquarius

This period of the Moon connects with Venus, easing tensions and lessening the creeping feeling of disenchantment you were prey to. You will feel, act and work much better than before. Health and dietary considerations must, however, be taken into account, warns Ganesha.

12 November: Full Moon in Taurus

You will be gearing up for better health, a more positive outlook, coping better with responsibilities and commitments, a bit of love and laughter, a certain amount of fun and games. Meets/conferences/functions may also keep you busy.

18 November: Moon's Last Quarter in Leo

Employment, loans and funds, the nitty-gritty of day-to-day life will be the theme now, along with public relations and group activities. Open rivalry, or certainly, competition will have to be dealt with. Marriage or business alliances and reaching out to people and places are equally strong indications.

26 November: New Moon in Sagittarius

You will be lion-hearted in combat this quarter, ready to go for the jugular, if so required. Parents and in-laws play an important role. You will be assertive and ambitious, yet will allow others to set the pace, knowing that you can match them, stride for stride.

4 December: Moon's First Quarter in Pisces

The Moon brings a realization of goals, a sense of achievement and also, strangely enough, a preoccupation with all things mechanical and electronic. I don't know how it will work out, but that is Ganesha's astrological message – your fascination for gizmos and gadgets, perhaps even surfing the Internet, inventing and discovering.

14 December: Full Moon in Gemini

This period brings nostalgia, memories, 'looking before and after', but you do not 'pine for what is not'. You are still in a mood for action, especially on the house/home/renovation/decoration front. Buying/selling and group interactions are also important

19 December: Moon's Last Quarter in Virgo

Sociability is your middle name of this week, and probably right through till the end of the month. You are innovative, and productive too, and in better shape physically and emotionally. Marriage, ties, bonding, children and even hobbies are all there to keep you on your toes. Sports, outings, coping with or developing new management techniques also interest you.

26 December: New Moon in Capricorn

You are both intuitive and artistic, along with a fine attention to detail verging on perfectionism. Concentrate on your creativity and/or to ignore and/or minimize tiffs, arguments, debates and angry words.

GEMINI

21 May–20 June

Gemini is the sign of the Inventor, the Artist. You Geminis are mentally energetic and inquiring, restless, artistic, fickle and contrary, witty, versatile, joyous and full of exuberance.

Ganesha says you Geminis are intelligent, versatile, witty, diplomatic, gossipy, and most certainly, very sharp on the uptake. The real tragedy is Donald Trump is also a Gemini but unlike you all he is insular, totally against migrants, slurs and maligns African migrants, has forcibly made Jerusalem the capital of Israel and hates all who oppose him. He is negative, a big bully and, most certainly, does not deserve to be the president of the greatest country in the world. I agree this is my opinion. I also agree that I could be wrong. But I have a right to what I think and feel and try to understand. I do not judge you Geminis. In other words, Trump is not the real and the ideal example of a true Gemini. He is abnormal.

This time I am following it up with your digital plan for the year so that you can focus very well on it in advance and thus prepare yourself by making a blueprint of the entire year. The digital plan for the year will be short and to the point, and though it is a general reading, it should apply quite a bit to you.

BLUEPRINT FOR THE ENTIRE YEAR

January: Joint finances, funds, loans, legacy, family problems.

February: Sweet and sour relationships, publicity, conferences and meets, inspirational and intuitive moves and manoeuvres.

March: Prestige, status, power struggle, perks, new ventures and means of communication.

April: Socializing, group activities, marriage, love affairs, happiness, laughter, the goodies of life.

May: Secret activities, health, expenses, visits to hospitals, welfare centres, medical check-ups.

June: Wish-fulfilment, happiness, money, marriage, confidence.

July: Money and honey, riches, beautification, augmentation of income, good food, jewellery.

August: Research, contacts, communication, correspondence, brothers, sisters, relatives.

September: Home, house, property, renovation, decoration, alteration.

October: Love, romance, children, relationships, hobbies.

November: Health, pets, servants, jobs, hygiene, colleagues.

December: Love, marriage, divorce, journeys, reaching out to people, also separations.

This time we are giving you extra information about your own sign. It will be useful forever and ever. Ganesha asks for a look at the decans. Ganesha says every sign of the zodiac covers 30 degrees which can be further divided into three decans of 10 degrees each, which can help narrow down your traits further, as they will vary even within the Gemini framework of 21 May to 20 June.

If you were born between 21 to 31 May (the first decan), the planetary ruler is Mercury and you often have exceptional mental brilliance. You are also very hospitable and dependable, having some Taurean traits as well. A good friend, full of ideas, yet practical. You are good at noble causes and the graphic arts.

If your date of birth is between 1 and 10 June (the second decan), Venus is the planetary ruler and your gentler qualities come to the fore. You love making people happy, and both nature and human beings hold a great fascination for you. You love beauty and harmony, but sometimes this softness in you tends to be exploited by others.

If you were born between 11 and 21 June (the third decan), both Saturn and Uranus rule over you, so that there is a certain toughness along with your essential creativity. You are capable of tremendous achievements, and yet retain a very human outlook. Psychiatry would be a suitable line too, though this is the decan of the inventor. The essential Gemini duality yields dividends in this decan.

Jupiter, the planet of plenty and prosperity, will be in your seventh angle from 9 March 2018 to 2 December 2019. Jupiter will be in your seventh angle, blessing you with a tour de force which means you will perform remarkable feats of work, and obviously, it will be quite an achievement. You have an umbilical cord–like relationship with love, romance, marriage, collaborations, ties, connections. Here are a few astrological tips which will be of use to you:

- Public relations and competition,
- Open rivalry and legal contests,
- Interest in change of locale,
- Marriage and other alliances,
- Accenting teamwork and understanding,
- Compromising major differences,
- Allowing others to set the pace, and
- Trying to be objective.

The key will be in cooperation, collaboration, teamwork, journeys. Right from 2014 there must have been a radical change in your lifestyle. And change must have been both personal and professional. This change will continue till October 2020. Therefore, it is the main trend which will alter your life.

Strictly speaking, the seventh angle stands for the whole world except your own self. Therefore, it means attachments, ties, knots, relationships. Normally the biggest relationship is marriage. At the same time the biggest relationship is also divorce. Therefore, the seventh angle stands for all legal matters too. The seventh angle is the most vast angle because it includes everybody else except yourself. The key is in the connection and relationships. So simple and so final. Many times it also includes journeys with a stopover. In other words, it is a halt between two destinations. Finally, the seventh angle speaks about your relationships with Nature, society and even God.

Saturn, the planet of duty, responsibility, limitations and the one balancing the scales of life itself, will be in your eighth angle from 21 December 2017 to 23 March 2020. The spin-off will be the total finance scene, comprising earned income, funds, loans, investments, legacy, joint finance and capital formation. It will, in many ways, make the trends and tendencies of last year run into this one. However, it's not only money but also the entertainment and communication world that keep you occupied – food, fads and fashions, family life – all the Fs, counting finance. Food could even extend to include dietary and nutritional concerns vis-`a-vis the household and domestic matters, which cannot really be relegated to second place, even though business is your major thrust. You'll just have to push that little bit extra.

Talking about food, let me give you an illustration of the largest fondue in the world. As you know, America is symbolized by the largest, and the largest fondue was prepared on 28 February 2007, under the supervision of chef Terrance Brennan, owner of the Artisanal Cheese Centre in New York. The fondue weighed 2,100 lbs (952.56 kg). Its major ingredients were 1,190 lbs (539.78 kg) of Gruyere cheese from Wisconsin, 120 lbs (54.4 kg) of white wine and some spices. It was prepared in a 220-year-old 2-tonne cast iron kettle brought from Louisiana.

The greatest lesson Saturn teaches you Geminis is health is wealth. You need to do exercise, build up your muscle and have an extra dose of strength and stamina. This is mighty important. Also taxes, joint finances and property, improving your self-confidence, trying to save wherever it is truly possible but not to be miserly. In short, a fine balance between lending and borrowing. Trusts, wills, dealings in property matters and insurance, and if you are old, preparing for a graceful retirement are a must for you Geminis because of the simple reason that you live on your nerves and sometimes take rash decisions and actions which might prove infavourable. I would say in short that care and caution will be better than enterprise and taking new challenges. But I have always maintained that astrology is not perfect and the final decisions is always yours. I can only suggest and comment to the best of my limited abilities.

Your happiness quota will be 81 per cent.

SPECIAL BONUS

Ganesha says my devotee Bejan is introducing the topic of special bonus for all of you. This special bonus is placed upon Mercury. Mercury represents journeys, ties, trips, mobility, migrations and all types of communication. Venus represents love, beauty, the fine arts, polish and finesse, diplomacy and persuasion, joy and a big hurrah to life itself. Therefore, I am combining both communication and joy for you and giving you the necessary dates.

For you witty, clever and sharp Geminis, Mercury is your main planet. Mercury, as you know, is the messenger and the contact man of the zodiac. Mercury turns favourable for you from 24 January to 9 February, 17 April to 6 May, 21 May to 4 June (very definitely), 27 June to 18 July (good time to combine romance and finance), 11 to 28 August (once again romance and finance), 14 September to 2 October (research, studies, journeys and a comprehensive view of life), 9 to 28 December (love and hate in mighty proportions).

Children, hobbies, romance, entertainment and, surprisingly enough, hospitalization and looking after the lonely and unfortunate all come under Venus for you. The periods will be 7 January to 3 February, 2 to 20 March (lucky and important), 27 March to 20 April (give your best shot in whatever you do), 21 April to 4 May (sorrow, happiness, expenses and profit all go together), 9 June to 3 July (great self-confidence and victory), 28 July to 20 August (you will be at your brilliant best), 15 September to 8 October (by your versatility and talent you will outpace and outrun the rest), 12 to 25 November (mixed results right from lawsuits to new romance and broken relationships), 20 December to 13 January 2020 (research, hobbies, widening your mental and emotional horizon. In short you will be evolving).

However, I must be true to my own self. This is only a solar-scopic reading based mainly upon the position of Sun. Therefore, it may not be as accurate as a personal horoscope. At the same time, I have put in a lot of effort, imagination and inspiration as well as intuition into it. In other words, I have tried my best. The results are up to our maker. Yes, sometimes miracles do happen. Therefore, I end on a positive note, though I know my limitations very well.

KEY TO WEALTH BY NASTUR DARUWALLA

Geminis can turn unfavourable conditions to favourable ones using their common sense. They are by nature stubborn and hard-working, and can easily overcome difficulties. The mantra for them is: *'Om Shrim Shriyai Namah'*. Donate shoes or clothes to children.

WEEKLY FORECAST BY PHASES OF THE MOON

6 January: New Moon in Capricorn

You start the year in a highly emotional, creative and sensitive frame of mind. Children, hobbies, even poetry to cookery – the range is enormous. Partnerships, collaborations, and material alliances are all there as the main thing is reaching out to people and places.

14 January: Moon's First Quarter in Aries

An overseas slant and the human angle claim you in this period. Journey, liaisons and a rendezvous are highly probable and the link is escalating expenses. You may reach out to help those sick in spirit, mind or body and you try to do it to the best of your ability, approves Ganesha.

21 January: Full Moon in Leo

January well prove to be a milestone in money matters. This is a trend ushered in by the Moon, with the focus on new avenues of earning, buying and selling, loans, funds and joint finances. Brothers, sisters, kin and neighbours draw close, and, family-wise, it is a time to strengthen existing bonds and forge new ones.

27 January: Moon's Last Quarter in Scorpio

You will make much headway. News, views, messages and calls are important and urgent, so also travel. You may well be able to consolidate your position, and some form of promotion, perks and advancement are very probable.

4 February: New Moon in Aquarius

The Moon makes you valiant and determined. Existing gains and the position you have attained are further consolidated. An additional gain is a developing predilection for yoga, the occult and meditation, and interest in salvation, regeneration and the hereafter.

12 February: Moon's First Quarter in Taurus

A good time to give or receive favours, to show generosity. Your personal and financial angle is strongly emphasized. Children, creativity, news and views are also given a great boost. So also do your emotions and libido receive a surcharge that electrifies them and you feel you have truly come alive.

19 February: Full Moon in Scorpio

The Moon triggers off finances and communication, two things which really influence destiny, whatever you may say. You will have to make split-second decisions, move fast. Much of your actions and activity now will have spin-offs in March and onwards. This is decision time, action time.

26 February: Moon's Last Quarter in Sagittarius

A little health care may prove necessary. Yoga and meditation will help. Your ambitious and hard-working nature will make you pull out all the stops and sometimes you may find you have overextended yourself. Beautification or renovation or acquisition of house/home/office/workplace are favoured.

6 March: New Moon in Pisces

The decisions taken earlier may take some effort to sustain. A tie may finally be broken but it's no use giving in to remorse and regrets. You will soon bounce back. This may not only be the start of a very strenuous and taxing time, but also highly exciting and strenuous.

14 March: Moon's First Quarter in Gemini

People are in strong focus this period. You extend your hospitality and heart, thus overcoming a slight feeling of melancholy and sadness. I'd like to remind you of the famous Spanish saying, *'Mi casa estu casa'*, and you will see how you reap dividends in terms of both popularity and authority.

21 March: Full Moon in Libra

The Moon makes you open out. Messages, news and views, relatives and visitors, all jostle for attention. Also, trips, links and ties. Most of your difficulties will soon be over, if not already so. You will now have learned to shrug or laugh philosophically and say, 'Business as usual.'

28 March: Moon's Last Quarter in Capricorn

This quarter of the Moon has you in a sentimental and nostalgic mood. A blending and blurring of the boundaries of past, present and future may make you feel you're inside Jules Verne's time machine. However, journeys, the domestic scene, rituals and children are the details of this quarter's forecast. Ganesha adds a bit of meditation too!

5 April: New Moon in Aries

Though you still feel slightly vulnerable, you also go through a very human, humane and lovable phase now. Activities involving superiors, parental ties and duties will benefit, so also business/profession, if you are responsible, diplomatic and on less of an ego trip.

12 April: Moon's First Quarter in Cancer

The angle this quarter is definitely the Cs of consolidation, creativity and children, the Ps of passion and power, and much socializing. Quite a formula! All ties and attachments, some easy money, go a long way towards lessening your woes. Ganesha nods.

18 April: Full Moon in Libra

Matters connected with taxes, insurance, rentals, leasing and joint finances will be important. Wining, dining, invitations and functions will claim your time and attention. Your sources of income and earning may also multiply. The financial situation will certainly be easier.

26 April: Moon's Last Quarter in Aquarius

The productivity and creativity ushered in by the Moon continues. The focus will be on your bank balance and credit, on home conditions – but diplomacy and bargaining skills will be required. There may well be promotions and/or additional perks in the offing, adds Ganesha.

4 May: New Moon in Taurus

This period will demonstrate how the efforts put in all through the period will pay now in this phase of the Moon. Ganesha claims that apples of gold could well fall into your lap. Your health will improve, and you give your best in your chosen field of activity, and there is a degree of satisfaction in that.

12 May: Moon's First Quarter in Leo

This period makes you rather restless, introspective and, at the same time, ingenious and inventive. You may well cover a lot of territory, even in lonely, out-of-the-way places. Spiritual pursuits, soul-searching, psychic and occult phenomena, all draw you inexorably as well.

18 May: Full Moon in Scorpio

A wish-fulfilment is predicted, in the name of Ganesha. A lucky phase that will last till June is starting now. Sports and holidaying, trips and friendship, artistic endeavours are all favoured. Tourism will be full of profit. The nitty-gritty details you'll have to work out for yourself.

26 May: Moon's Last Quarter in Pisces

Socializing, a sense of well-being, improved finances come your way this quarter. Along with them, of course, come added responsibilities. They always do, winks Ganesha. A marriage or other alliance, collaborations and partnerships, or a position of responsibility and trust are also foretold.

3 June: New Moon in Gemini

This quarter sees the culmination of the trends of the last period. All the media of communication and entertainment will be used to your advantage. You will not only be in the right place at the right time, but meet the right people to make things happen to your advantage. Wonderful!

10 June: Moon's First Quarter in Virgo

An introspective and restless slant to this period. Paradoxically, while you look inwards you also look outward at foreign connections and a fair amount of travel. Emotional extremes and disregard for others are to be avoided like the plague.

17 June: Full Moon in Sagittarius

Augmentation of money from all kinds of sources. Also of your psychic and spiritual resources. Loans, funds, even a lottery or games of chance have happy outcomes. Passion and pleasure also add their bit to make this period, and you, truly dynamic and exciting, Ganesha be praised!

25 June: Moon's Last Quarter in Aries

Channels of communication, partnerships, travel, rendezvous, and relationships are the focus in this period. Help, when and where you can least expect it, will materialize. Amusements and entertainment will keep you busy, perhaps hectically so.

2 July: New Moon in Cancer

Bizzare but beautiful people and situations swim into your ken. Wills, legacies, probate and so on, if not already initiated in this period, will be important. A practical, all-out, no-nonsense approach to making money will work. After all, money does make the mare go!

9 Jul: Moon's First Quarter in Libra

This quarter helps you clear away all dead wood, fix your sights firmly on aims, objectives and personal goals. Promotions may come, but will bring heavy, even burdensome responsibilities. Financial matters and personal belongings and possessions are also important. Reserve funds must *not* be neglected.

18 July: Full Moon in Capricorn

The trend for the month includes all the Fs of finance, family affairs, even fashion, food and fads. You are bold and charismatic in your attitude, and this may prove to be the ideal time to push things through, or start a new enterprise.

25 July: Moon's Last Quarter in Taurus

You are as sensitive as can be, to life itself. Personal progress is made, but also on the work front. You acquire a degree of control over your wild impulses and develop a keener cutting edge vis-à-vis career, skills and people. That's saying a lot. But Ganesha says it.

1 August: New Moon in Leo

Moon madness – a delightful feeling – could well possess you in this period. Don't let it overwhelm you, though, and keep your feet firmly planted on terra firma. Being romantic and playful is good upto a point. Music, creative pursuits, even sexual liaisons are highlighted, also sports, hobbies, speculation.

7 August: Moon's First Quarter in Scorpio

An ideal period to be forward and outgoing. The focus will be the social whirligig, contacts, communication, and creative pursuits – the three Cs once again. An ideal time to accomplish many incomplete or abandoned things. Also, trade connected with art and artefacts and export may get an added boost.

15 August: Full Moon in Aquarius

The theme for August is definitely productivity, ushered in this quarter of the new Moon. The month and sign tally. In many ways, it is the deciding week for you, but the camera zooms in on financial gains and emotional satisfaction. Loans and funds will work out and friends will not only socialize, but stand by you too.

30 August: New Moon in Virgo

Everything concerning you gets personalized in this period; you are very emotional, intense and withdrawn, even secretive. It is time to cleanse and uplift your heart, soul and spirit. You now gain an inner strength that helps you get things done in what in yuppie language is referred to as 'the real world'.

6 September: Moon's First Quarter in Sagittarius

The Moon endows you with artistic excellence, a sense of fun, joy from offspring. You have the foresight and wisdom, also patience, to untangle problems. Love is a strong motivating and healing factor for you and that is something to be eternally grateful for, always and especially in today's world.

14 September: Full Moon in Pisces

This period brings you gains from buying and selling, brokerage and trade. This will also extend to include sale/purchase and so on of house and home, properties and/or office space. This may include also comforts, luxuries and necessities for better living.

22 September: Moon's Last Quarter in Gemini

The Moon ushers in the *action theme*. An exceptionally productive and lucrative phase of your life will have repercussions on journey, finance, marriage, the world of advertising, and print and electronic media too. Letters, calls, faxes – communication is extremely important right now.

28 September: New Moon in Libra

The influence and impact is now on your angle of work, with gains foretold. Awards, rewards, promotions. Easy money including even from unexpected sources is more than probable. Cash, kind, future benefits are handed out by Ganesha.

5 October: Moon's First Quarter in Capricorn

The Moon spurs you on to hectic, even manic activity. Entertainment, research, products of the imagination, computer technology and, powered at full throttle, you display rare qualities of leadership and enterprise

13 October: Full Moon in Aries

This quarter is for you to win friends and influence people with the minimum of effort. Dale Carnegie could take lessons from you. Children, consumer goods, cookery are the three Cs now, along with family, entertainment, machines and gadgets, and just good, clean fun.

21 October: Moon's Last Quarter in Cancer

The Moon brings some contrary, if not contrasting, pulls. Mounting expenses, secret and superstitious activities, travel and yet an inexplicable discontent or disenchantment. It's not that you won't work, or not work hard, but it's as if you find something missing.

28 October: New Moon in Scorpio

You feel you've been given a second chance to make good. And how many of us are so blessed, asks Ganesha's devotee. Personal affairs may give you a rough time. This is not the time to be sensitive and sentimental. And in any case a little pragmatism is necessary to get through life.

4 November: Moon's First Quarter in Aquarius

You are hurled willy-nilly into the social whirligig. Strong and intense friendships may develop, new sources of income also. Team work will stand you in good stead. You will see results, and that too pretty fast.

November 12: Full Moon in Taurus

'Be firm of purpose,' says a Shakespearean character, and that could well be Ganesha's message to you for this period. You are given a second chance and are extremely likely to come out on top. But remember another saying, 'Fortune favours the brave,' and that applies to you at this point.

18 November: Moon's Last Quarter in Leo

You are launched, in fact sky-rocketed, into the big-time league. Contacts and communication is where the action is. All kinds of activities, in fact the entire gamut of them, will make demands on you. And you rise to the challenge, beautifully so.

26 November: New Moon in Sagittarius

The contacts and communication trend of the last quarter carries over into this one too. It is, in fact the monthly trend. Those dealing in the spoken and written form of communication, writers, admen to newscasters, to name a few, stand to gain more. Trading and speculation are also favoured. Same as in Aries.

4 December: Moon's First Quarter in Pisces

The Moon sees you totally fascinated by inventions, machines, electronic devices, the entire range of the latest gizmos. It is a fascination that will last for some time and even prove to be lucrative in some respects. Your skills and innovative strategies stand you in good stead.

14 December: Full Moon in Gemini

Pets and projects, the job scene, loans, funds and dependents, including pets, employees and servants are the list that Ganesha gives me as your main concern. Entertaining and food will be highlighted and also some kind of health issues, for yourself and for loved ones.

19 December: Moon's Last Quarter in Virgo

The focus is definitely on personal matters like marriage, or marital concerns, problems and funds, or those pertaining to children. You are at the centre of it all, and also of meets, functions and conferences. I must point out that you give more than a good account of yourself, by the grace of Ganesha

26 December: New Moon in Capricorn

Arguments, tiffs and words spoken in anger are likely this week. It is a part of life, but anger and acrimony have to be curbed. You will be artistic, intuitive and creative, and you will find the sun shining and the storm over. The Moon influences the marriage angle, and give and take are what makes a marriage work, and last, especially in these troubled times we live in.

CANCER

21 June–22 July

Cancer is the sign of the Teacher, the Prophet. You Cancerians are sensitive and sympathetic, moody, tenacious, ambitious, caring, mediumistic.

Service to mankind is service to God

—Swami Vivekananda

To me the ideal example of a true Cancerian is the Dalai Lama. I may be right or wrong but at some level compassion meets supreme consciousness.

Ganesha says you get down to hard work and are disciplined. You want to make up for lost time. Your critical faculties are sharp. You realize that you have to brush up on interpersonal skills, as you often communicate what you don't intend to. Despite this, you sign big deals and succeed in important negotiations. You still manage to make a positive impression and earn the kudos of your peer group. You may also take to singing, poetry, painting and writing. You have many creative skills and they are also stress busters. There could be festivities and celebrations at home, marriages and engagements. You may even have an addition to the family, or bring home an exotic pet, like an African parakeet. Life is calling and you are ready for the dance. Ganesha journeys with you.

This time I am following it up with your digital plan for the year so that you can focus very well on it in advance and thus prepare yourself by making a blueprint of the entire year. The digital plan for the year will be short and to the point, and though it is a general reading, it should apply quite a bit to you.

BLUEPRINT FOR THE ENTIRE YEAR

January: Marriage, ties, love, collaborations, romance, meeting and reaching out to people and places.

February: Health, funds, tantra and mantra, change of locale, moving.

March: Journeys, publicity, ceremonies, collaborations, functions, rites and religion.

April: Stepping up on efficiency, work, status, prestige, taking care of parents, elders.

May: Help, socializing, friendship, fraternity, camaraderie.

June: Expenses, losses, spirituality, helping others, charity, long-distance connections.

July: Power, perks, promotions, prosperity.

August: Finances and family.

September: Contacts and joy.

October: Property, parents, in-laws.

November: Joy, creativity, children, hobbies (you make news and win over others).

December: Work, funds, employment, health and medical check-ups, servants, subordinates.

This time we are giving you extra information about your own sign. It will be useful forever and ever. Ganesha asks for a look at the decans. Ganesha says each zodiac sign extends across 30 degrees, of the total of 360 degrees. These, divided further into three parts of 10 degrees each, form the decans. Though all those born between 21 June and 22 July come under the sign of Cancer, the decans yield further insight into your total personality.

If you were born between 21 June and 1 July (the first decan), you are ruled by the Moon and the pure Cancerian traits are very pronounced in you. Tenacious, sympathetic, loyal and idealistic, you can be moody, suspicious, pessimistic and vacillating, a nagger with an inferiority complex. Try to build up more on the plus side.

If you were born between 2 and 11 July (the second decan), your ruler is Mars, with all its strength and vigour which it imparts to you. You are a stronger Cancerian, tending to almost a dictatorial and domineering temperament. You have tremendous energy and 'go' and will gravitate towards the limelight and publicity.

If you were born between 12 and 22 July (the third decan), you are ruled by Jupiter, the great benefic, and are sure of fame, good fortune and plenty of money. You are born lucky, except for an unfortunate tendency to fat! You benefit greatly from education and travel.

Now to the nitty-gritty. Right from 9 March 2018 to 2 December 2019, Jupiter, the great benefic, will be in your sixth angle focusing on the following:

- Health and relationships with colleagues, subordinates, servants, relatives, and here, you are advised by Ganesha to stay cool.

- Financial muddles are possible, but you will pull through, because of Jupiter's blessings, which I will expand on later, but better remember that loans, and funds, buying/selling, borrowing and lending will have their sharp edges and difficulties.

- Danger of theft, pilferage, misplacement of documents, valuables and money; better take care.

- Problems posed by pets, dependents, relatives, maybe because of circumstances beyond their immediate control.

- Stiff competition, for which you should better work systematically and give your best.

- The need to eat intelligently and serve others, at the same time taking enough rest and cultivating moderation in all that you do.

Saturn will be in your seventh angle from 21 December 2017 to 23 March 2020. The results will be bitter-sweet. Why? Saturn will be in direct opposition to your Sun. In simple words, it means there will be a tussle between Sun (power) and Saturn (responsibility). Therefore, the only way out for you is to balance these two. The only way to do it is by compromise, diplomacy and conciliation. In other words, Saturn will blow hot and cold over you. This could disturb you enormously if you do not keep your cool, your poise, your equilibrium and your balance. If you believe in mantras the specific mantra for you is: '*Hum hanumate ramadutaya namah*'. The other mantra is: '*Ram charana sukha payo*'. The first mantra indicates that Rama controls Hanuman. The second one says that you will find peace and relief by worshipping the feet of Rama. Modern people might find this stupid and superstitious. They are welcome to their belief. But I am a devotee of Hanuman, Rama and Amba Mata. My faith is my strength and my armour. Therefore I believe in it. Belief becomes a certainty and a guarantee if you believe sincerely and this belief should be in your blood and bone and even in your marrow. This is my opinion. At the same time, I respect your right to have an opinion altogether different.

Saturn could lead to legal battles, sorrows, separations and sometimes the possibility of death which is the final separation. I admit I do not have all the answers. But this much I know that Saturn is duty. Duty is beauty to me. Therefore, to me the answer is do your duty and leave the rest to your own maker.

Saturn in your seventh angle could sometimes lead to retirement or taking up a different set of values, a change of environment, competition. Be cooperative, try to understand the other person's point of view, and please do not think that only you are clever, intelligent and wise. Please understand, my dear Cancerians that in this world many people know all the tricks of manipulation and mischief and how to get work done from others. Therefore, you might face a tricky situation. Dalai Lama the Cancerian is my hero. Yes, I have met him.

What is my final answer to all this? It was the great storyteller, doctor, spy, Somerset Maugham, who said,

'Goodness is its own reward.' This might be difficult for you to swallow. It could be a bitter pill but that, my dear readers, is my final answer. I am now eighty-seven. The world is a grand mix of goodness, technicalities, spiritualities, wickedness and sorrow as well as misery and evil. We human beings are complicated and complex. We have it in us to be God's or to the Devil's The choice is yours.

Your happiness quota will be 83 per cent.

SPECIAL BONUS

Ganesha says my devotee Bejan is introducing the topic of special bonus for all of you. This special bonus is placed upon Mercury. Mercury represents journeys, ties, trips, mobility, migrations and all types of communication. Venus represents love, beauty, the fine arts, polish and finesse, diplomacy and persuasion, joy and a big hurrah to life itself. Therefore, I am combining both communication and joy for you and giving you the necessary dates

Mercury helps you from 10 February to 16 April, 27 May to 20 June, 25 June to 26 August (most important), 19 July to 10 August (very important) and 3 October to 6 December.

Venus helps you from 27 March to 20 April, 15 May to 8 June, 4 July to 27 July (most certainly), 21 August to 14 August and 9 October to 1 November (a fine period).

However, I must be true to my own self. This is only a solar-scopic reading based mainly upon the position of Sun. Therefore, it may not be as accurate as a personal horoscope. At the same time, I have put in a lot of effort, imagination and inspiration as well as intuition into it. In other words,

I have tried my best. The results are up to our maker. Yes, sometimes miracles do happen. Therefore, I end on a positive note, though I know my limitations very well.

KEY TO WEALTH BY NASTUR DARUWALLA

You are known for your moodiness, you are quick to be upbeat but just as quickly you are in a black hole. For you the mantra is: '*Om shri mahalakshmyai cha vidmahe vishnupatnyai cha dhimahi tanno lakshmi prachodayat om.*' Donate school bags or books.

WEEKLY FORECAST BY PHASES OF THE MOON

6 January: New Moon in Capricorn

'A call to arms' is the opening note of the year, but your sense of duty is directed firmly to the home and family angle. There is a personal touch to all that you feel and do, strongly driven by your intuition, imagination and emotions, which govern all your participation in all activities/events at home or at work.

14 January: Moon's First Quarter in Aries

Travel and news from a distant place – even from foreign parts. They will be a source of both pleasure and encouragement to you. Journeys will be profitable as well. I am risking my professional reputation by telling you to follow your heart. That's the way you'll get the best results in almost all your endeavours.

21 January: Full Moon in Leo

This phase of the Moon makes you a scene stealer, especially your public image. Action and attachments – these two sum up the theme for the month. A personal attachment formed now could easily lead to a wedding, or a long-term commitment. Bully for you, chortles Ganesha's devotee. Once again, news from a foreign land is highlighted.

27 January: Moon's Last Quarter in Scorpio

The trend described in the previous quarter will still prevail during this period. Journeys and news are still highlighted. Also important will be group activities, your image in the eyes of others and reaching out to people along with places.

4 February: New Moon in Aquarius

This quarter sees you entertaining and being entertained. Hospitality – yours and their's – is the focus. You may be the star attraction at get-togethers and meetings. Whatever it may be, you are certainly in the limelight. Thus, it is your 'people skills' that count most. Money comes alongside.

12 February: Moon's First Quarter in Taurus

You have the energy and drive to get things done. Sensual and sexual urges may also be activated. Otherwise, this energy helps you to overcome low vitality, illness, a feeling of malaise. Inspiration and innovativeness may also register an upswing.

19 February: Full Moon in Scorpio

The Moon makes for a strong focus on money – better income, sudden gains (legacies, lotteries and such), lending and borrowing, hypothecation and brokerage come within its ambit. You may be handling large sums of money, public funds and trusts and finances.

26 February: Moon's Last Quarter in Sagittarius

Emotions colour your actions during this quarter. Children and leisure pursuits are a source of joy. You have the Midas touch in what you see, do, handle – without the risk factor. Your luck holds, and your charm works miracles. The Midas touch, of course, gives you more money.

6 March: New Moon in Pisces

These are useful and exciting days for legal and official matters, contacts and communications. Team work, understanding, cooperation – these are important and necessary. Alliances too are important, whether marriage, business or something else.

14 March: Moon's First Quarter in Gemini

Trips, ties, relations with neighbours and concern with your environment gain precedence now. Dealings with brothers and sisters are also included. Likewise, for the use of media, advancement of learning and knowledge, and even a knowledge of artificial intelligence.

21 March: Full Moon in Libra

A culmination point in your evolution may have been reached. This is what the Moon brings you and you cannot rest on your laurels as opportunities of a lifetime will materialize. Leave time for rest and relaxation and learn to space yourself out, or you'll never cope with what lies ahead. This is only the beginning of a new phase in your life, says Ganesha.

28 March: Moon's Last Quarter in Capricorn

News, views, messages, relatives and in-laws, entertaining and travel. You may be justified in asking with heavy sarcasm, 'Anything else?' But that's very definitely the astrological message. You'll feel you can never do enough to cope.

5 April: New Moon in Aries

The house/home/office will be strongly in focus. Those in employment may get a superb, genuine offer – not one that you cannot refuse. Help from unexpected quarters is another strong likelihood. Your creature comforts – even luxuries – get a tremendous boost.

12 April: Moon's First Quarter in Cancer

The partnership angle will be in focus now. Human relationships will be your forte. What you do now may well have a long-reaching effect which will be more visible in May. A fair amount of social interaction, and networking on the one hand and also a separation or parting on the other are likely.

18 April: Full Moon in Libra

All your activities get a great push forward this quarter, and the momentum will not even give you time to catch your breath. Work will predominate but efficient and diplomatic handling of domestic affairs, of parents, in-laws, even your boss will be necessary. The Moon makes you rise to the occasion.

26 April: Moon's Last Quarter in Aquarius

A fine period for all of the above, and for a very real and intoxicating taste of power and wealth. The two are usually interrelated, aren't they, in today's materialistic world?

4 May: New Moon in Taurus

A spiritual journey, initiated in March, has been running underground even as you've been hectically busy with mundane matters. It now appears in a sense of duty, responsibility and trustworthiness, colouring all your activities and commitments. This will encompass both work and family and the larger or total environment.

12 May: Moon's First Quarter in Leo

You may tend to bite off more than you can chew, to attempt impossible feats. Rationality and restraint will be not just desirable, but necessary. The reason is, attachments and ties as well as entertainment, pleasurable socializing and group activities are highlighted.

18 May: Full Moon in Scorpio

This phase of the Moon ushers in a period of ease, relaxation and respite from worry. Money will come very easily, even from sources other than hard work. Love and romance will give you happiness, and so will children. This is a fine trend for anyone, and Ganesha has been truly generous. Or is it Fate? I prefer to think the former.

26 May: Moon's Last Quarter in Pisces

Your intuition is highly developed and an accurate guide for you. This is the time to make decisions that are vital, make fresh or difficult choices, in short, take the plunge. Loans and funds are favoured, so capital formation will not be problematic. A good time for investments too. Immigration, shifts of office or home are also generally favoured.

3 June: New Moon in Gemini

The highlights, besides what have been mentioned earlier will now be associations and alliances (even weddings), the birth of, or joy from, children and a lot of partying and eating out! You've really got it made for you this month. A really fine month, by the grace of Ganesha.

10 June: Moon's First Quarter in Virgo

The fun and games of last month, though not entirely over, certainly have to be toned down. The emphasis is on family and unequivocally on work. Practicality and common sense will be necessary and you will have to think about providing for the future in all you do and plan. You will work with a will – Cancerians always do.

17 June: Full Moon in Sagittarius

This month it's a trend of mixed blessings. The work theme has already been mentioned. If you are thinking of shifting home, a change of locale, migrating to foreign parts, unexpected hurdles and hindrances will come up. Some regards and recriminations are likely. One thing is for sure, though. You will be meeting a lot of people, having a lot of visitors.

25 June: Moon's Last Quarter in Aries

A home away from home is possible, but the focus on home can take the form of renovation, decoration, doing up and beautification too. You will have to get smart in this month and be practical to the point of ruthlessness, in order to succeed in your endeavours.

2 July: New Moon in Cancer

While house, home, travel and family continue to be focused, the Moon endows you with astonishing artistic ability and talent, particularly in the field of music and dramatics (a Cancerian strong point), if so inclined. Working from home or combining home and office are also likely.

9 July: Moon's First Quarter in Libra

This is decision time. Actually this was ushered in the previous quarter. Partnerships (whether in marriage/ profession/business), starting a new venture, or moving house or migrating – these are all definite possibilities. At the same time your own inner well-being, as well as health are important. Compromises, giving in, making adjustments will give inner peace and happiness.

18 July: Full Moon in Capricorn

Work and material things once again take centre stage. A definite feature will be *change*. You may have to change or alter work habits or plans. The reason could well be concern about elders, relatives or parents. Whatever it is, you will certainly be putting in very long hours of work and find your schedule disturbed and your health affected. Ganesha's advice: rest, unwind, keep calm. This will be the monthly trend for July. So I suggest you heed it.

25 July: Moon's Last Quarter in Taurus

You will be sensitive to a fine degree, in tune with life itself. The impetus or push given in the previous phase will sustain itself, making you keen, efficient and skilful at work. In addition, you will strive for, and achieve, a degree of self-control. No more daydreaming and vacillating for you. You therefore make great personal progress.

1 August: New Moon in Leo

Children, hobbies, creative pursuits and sports and speculation are all highlighted. Fun and frolic, passion and sex, music and money matter too. Lots to do, and in addition you feel charged with vitality and vigour, the ability to do it all. It frequently happens in the birth month of a sign, and July is that for you Cancerians.

7 August: Moon's First Quarter in Scorpio

In several respects, a continuation of the trends and activities of July will be seen in August. Socializing, group activities, hosting and guesting (to coin a phrase), money and marriage are the way it goes. It's still time to rejoice in your good fortune and be generous to others.

15 August: Full Moon in Aquarius

This could be the deciding quarter for you, in many ways. Financial benefits and emotional satisfaction are foretold. An attachment or even engagement could lead to marriage. Friends give joy and loans; funds, finances will be organized swiftly. Go all out for what you want – that is Ganesha's advice. The Moon ushers in a long and happy ride to success and achievement.

30 August – New Moon in Virgo

Loans, funds, group activities, social functions are all in focus. It is a culmination of the trend started in July, in almost all respects. Your energy and enterprise will still be sustaining itself, so you will be helping the planets in making things happen for you.

6 September: Moon's First Quarter in Sagittarius

In this quarter of the Moon the theme will be communications and messages. Travel is very likely; keep your travel bags packed and ready. Also, you may get help from both known and unknown sources. It is, paradoxically, also time to 'get real' and consolidate your position and existing gains.

14 September: Full Moon in Pisces

The Moon brings in the theme of communication, widened to include travel, publicity, media (both print and electronic) and transport. It may also bring a wish-fulfilment or the realization of a dream. Success in a project/venture/enterprise too. If you start the process now, important documents or contracts could be signed in November. Don't sit back and wait for my forecast to come true, says Ganesha. Get moving.

22 September: Moon's Last Quarter in Gemini

Your work angle is awards and rewards, recognition and/or promotion. A bonus, larger pay packet, raise or gain from unexpected sources could be greatly welcome fallouts from this. Future benefits are also highlighted, but work hard you must, to deserve and derive them.

28 September: New Moon in Libra

The Moon propels you almost forcibly into the public eye. You will be in the limelight, whether you like it or not, so you might as well learn how to stay there. The right time for the right thing, says Ganesha. You may have to do a bit of a hard sell on yourself. It'll get you places.

5 October: Moon's First Quarter in Capricorn

House, home and personal affairs will be the focus now. Safeguard your health, don't overdo things even though the last few months (July onwards, in fact) have been truly action-packed and have had you in a spin. You may be tempted to draw into your shell to restore your equilibrium, but that's not advisable.

13 October: Full Moon in Aries

You shine at personal relationships, get love and friendship and appreciation. Renovation, decoration, beautification of the home, perhaps even a house-warming party, or acquisition of land or property are foretold. Also, a kind of wilting under pressure, of feeling raw at the edges, or highly strung. This is the monthly trend, so you really can't crumble under pressure, warns Ganesha.

October 21: Moon's Last Quarter in Cancer

This quarter may focus on a strong bond, relationship or personal attachment which may well lead to marriage in early 2019, if so situated. Those already married will experience a degree of warmth and rejuvenation that puts the magic back. Alliances and partnerships of all kinds are highlighted.

28 October: New Moon in Scorpio

The Moon will see the trend of the previous week widen and focus more on the home and hearth. You may either rebuild or renovate or even think seriously of a change of ambience and environment. You may embark on a journey of rediscovery to retrace your antecedents, childhood memories and roots. At the same time, a new project may be started even as some pending matters are concluded.

4 November: Moon's First Quarter in Aquarius

Into the house-and-home theme will come family, kin, in-laws and neighbours. Avoid making changes for the sake of change. Sometimes, if you disturb the status quo, unnecessary troubles and showdowns are likely, with unhappy outcomes. You will have to use a fine sense of balance and judgement.

12 November: Full Moon in Taurus

Loans, funds, additional income from different or even unusual sources are this quarter's gifts to you. Also important will be shopping, acquisitions, buying, selling, even trading. Media-connected people, writers, copywriters do well and gain more. Results will begin to show, with the trend started with the previous quarter; very soon, good things will happen, and you will feel doubly blessed.

18 November: Moon's Last Quarter in Leo

You will find this month a banquet to life itself, in continuation of the previous quarter. You will be in the fast lane of life where all the action is. To spell it out – children, hobbies, creative pursuits, collaborations, happiness

CANCER

in relationships. Also lucky for romance, ventures and adventures. Ganesha has surely been most generous.

26 November: New Moon in Sagittarius

Wish-fulfilment and enthusiasm are the Moon's offerings. However, a word of caution from Ganesha: overreacting, fantasizing, overextending yourself can all have unhappy fallouts and nullify the good effects of the right words and deeds of this phase.

4 December: Moon's First Quarter in Pisces

The lucky areas for you will be those encompassing the angles of art and beauty, home and house, work and status and, finally, love. How well and wisely you encash on them, how you make them work for you, depends on you and you *alone*. This is the message of Ganesha while handing over to you that extra dollop of luck.

14 December: Full Moon in Gemini

This quarter of the Moon is in your angle of valour, daring and bravery, trips and adventure, ties and relationship. Also research, study, the pursuit of knowledge and intellectual stimulation. That's saying a lot, but Ganesha says it. Also, please read the concluding bit of the preceding quarter. You'll know what to do with it. This is for the whole month.

19 December: Moon's Last Quarter in Virgo

This quarter of the Moon, and over the month and the years, the trend prevails over them all and it is strongly and uncompromisingly on *work* – please note the emphasis. It is what you must capitalize on. The heavy pun is to catch your attention. A promotion, change of job, moving on to fresh pastures are all likely.

26 December: New Moon in Capricorn

Everything gets personalized for you this quarter and carries you into the next year. A dream comes true; a goal is realized. You may find yourself irritable and edgy but, on the whole, keep in what has been, with Ganesha's blessings, a good year.

LEO

23 July–22 August

Leo is the sign of the King, the Lover. You Leos are arrogant, proud, domineering and self-confident, trusting, impulsive, generous to a fault. Boastful, but very brave. Self-awareness and ego are also seen.

Ganesha devotee Bejan Daruwalla laughingly poetizes it:

> Love is a ring, a zing, tantalizing
> Of love if it is good to sing.

Why this emphasis on love? We will come to the reason later on, but let me suggest that it is Jupiter who is mainly responsible for it.

This time I am following it up with your digital plan for the year so that you can focus very well on it in advance and thus prepare yourself by making a blueprint of the entire year. The digital plan for the year will be short and to the point, and though it is a general reading, it should apply quite a bit to you.

BLUEPRINT FOR THE ENTIRE YEAR

January: Loans, funds, joint finances, domestic matters, job, health.

February: Love, hate, marriage, divorce, contradictory influence.

March: Loans and funds, health and pets, religion, spirituality, rites for the living and the dead.

April: Freedom, intuition, inspiration, publicity, long-distance connections.

May: Work, parents, status, rivalry, prestige, tremendous pressures.

June: Friendship, wish-fulfilment, material gains, socializing, group activities, happiness and health. You end on a positive, winning, winsome note.

July: Expenses, secret deals, negotiations, trips and ties.

August: Success, projects, ventures, funds, children, creativity, good luck.

September: Money, family, promises, promotions, perks.

October: Contacts, communication, contracts, research, import and export.

November: Home, house, renovation, buying/selling, ill health, retirement.

December: Fine performances all round; you strike it rich, are lucky and win applause.

This time we are giving you extra information about your own sign. It will be useful forever and ever. Ganesha asks for a look at the decans.

Ganesha says each sign occupies 30 degrees of the zodiac. Further divided into three decans of 10 degrees each, this can help to narrow down and pinpoint your traits with greater exactitude, though all those born between 23 July and 22 August comes under Leo.

If you were born between 23 July and 2 August (the first decan), you are ruled by Sun, and are destined for fame and success, since this planet gives you great reserves of vitality and tremendous enthusiasm. You are artistic and kind-hearted, and quite conventional. You are very dignified and stately, almost verging on pompousness. You have remarkable reserve and noteworthy abilities to take you far in life.

If you were born between 3 and 13 August (the second decan), your ruling planet is Jupiter. This is particularly favourable for material gains, wealth, travel, social status and position. On the other hand, you have religious propensities and also intuitive faculties almost bordering on extrasensory perceptions. You will almost definitely have epicurean tendencies, loving the good things of life, and will be surrounded by gaiety and cheer. Banking may also do justice to your organizing abilities.

If you were born between 14 and 22 August (the third decan), Mars is your planet. Your nature is intrepid, fiery, independent, aggressive and impulsive. Your self-confidence, daring and boldness verge almost on bravado. You resent any kind of criticism or opposition, no matter how well meant. You have the great Leo capacity for reacting to fresh stimuli and experiences. Take care of your health, for if you fall ill, it is likely to be critical.

Jupiter, the planet of plenty, will be in your fifth angle from 9 March 2018 to 2 December 2019. Jupiter, your planetary indicator by the old system of astrology, will be in the ideal placing for creativity, research, children, sports and hobbies, and the luck of the draw – and that, perhaps, is the most important of all. Entertainment and romance, drama and films, anything to do with computers, automation, graphics, designing, space, rockets, science, speed, gadgetry and gizmos and architecture will be your food and fancy! Yes, you will excel at it.

To me the Parsees are Jupiterians and their Lord is Ahura Mazda. Therefore I am quoting the love and spirituality of Jupiter in the fifth angle and combining it with Ahura Mazda:

> Ahura Mazda's First Thought
> Blazed into myriads of sparks of light
> And filled the entire heavens.
> He himself, in his wisdom,
> Is the creator of truth which
> Upholds his supreme mind.
> O Ahura Mazda,
> You who are eternally the same,
> Further these powers through your Truth.
>
> —*Gathas of Zarathustra*
> tr. By PILOO NANNAVATY

Saturn is in a sojourn in your sixth angle of health and work from 21 December 2017 to 23 March 2020. Yes, it does mean that responsibilities will be piled upon you with avalanche ferocity. But an avalanche does not happen every day! Neither does it mean that you will be completely overwhelmed or submerged under it. It only means you will have to work hard and be reasonably regular in all that you do and undertake. It will be difficult to play truant. You will have to be in good health to face challenges and opportunities.

This is really a good way to get going, says Ganesha. You've decided to leap headlong into action, chasing success and achievement, pleasure and joy – and this gives you the opportunity to do so in both love and money. Exciting new opportunities for gain and progress come in both spheres. Finances are handled well but it is your career/profession that you handle with inspiration and insight. Coupled with your customary daring, it leads most definitely to new openings, perks, benefits and new projects with a high degree of satisfaction. You may be ready to take some calculated risks too, knowing that you have the guts to see them through and ensure that they will pay off spectacularly. It's a good phase at home too; almost certainly a time of no romantic/sexual conflicts. If they do arise, you see to it that you resolve them peacefully, affectionately, and in a very proactive manner. That is exactly what your brave, action-oriented nature demands. Maybe it's your new-year resolution, but the results are spectacular. You have decided to go far, and you need the stamina. Unnecessary stress or even too much anxiety is bad.

As Saturn will be in your sixth angle from 21 December 2017 to 23 March 2020 let me sum it up for you by informing you to take care of your health, your diet, your relationship with servants, employees, pets and dependents. Finally, it is my observation that you Leos often demand too much from life and your companions as well as fellow workers. Please do not push the envelope as the Americans say. Also for elderly Leos it could well be a time for retirement. If this happens, do so gracefully and try to get another part-time job only if you are up to it.

Your happiness quota will be 84 per cent.

SPECIAL BONUS

Ganesha says my devotee Bejan is introducing the topic of special bonus for all of you. This special bonus is placed upon Mercury. Mercury represents journeys, ties, trips, mobility, migrations and all types of communication. Venus represents love, beauty, the fine arts, polish and finesse, diplomacy and persuasion, joy and a big hurrah to life itself. Therefore, I am combining both communication and joy for you and giving you the necessary dates.

Mercury the communicator and messenger of the zodiac will help you from 17 April to 6 May (very particularly), 21 May to 14 June (most certainly), 27 June to 18 July, 11 to 28 August (perhaps the best of all), 14 September to 2 October and finally 9 to 28 December.

Venus, for love, joy and entertainment, children and hobbies, pleasure and power, will help you from 7 January to 3 February, 21 April to 14 May (certainly), 9 June to 3 July (good time), 28 July to 20 August (a time of rollicking fun and frolic), 15 September to 8 October and finally 2 to 29 November.

However, I must be true to my own self. This is only a solar-scopic reading based mainly upon the position of Sun. Therefore, it may not be as accurate as a personal horoscope. At the same time, I have put in a lot of effort, imagination and inspiration as well as intuition into it. In other words, I have tried my best. The results are up to our maker. Yes, sometimes miracles do happen. Therefore, I end on a positive note, though I know my limitations very well.

KEY TO WEALTH BY NASTUR DARUWALLA

Leos are fearless, egoistic and courageous; they stand against injustice and this trait gets them a lot of fame among friends and relatives. For them the mantra is: *'Om shrim mahalakshmyai namah'*. Donate foodgrains and sugar to the needy.

WEEKLY FORECAST BY PHASES OF THE MOON

6 January: New Moon in Capricorn

You will start the year spending heavily on the domestic front of home and family. A trip or move may be mooted. There may also be a considerable degree of introspection and psychic impulses that are more than genuine. Collaborations and foreign links are highlighted.

14 January: Moon's First Quarter in Aries

People and places are important, whether for work or for emotional relationships. People in public relations, the arts, research, and sales and commissions will do more than well this quarter. A paradoxical situation may bring either an engagement or wedding, or controversy, a separation, break-up or divorce. Such is life, says Ganesha.

21 January: Full Moon in Leo

You will think big, act with immense chutzpah and intelligence and foresight. These gifts from Ganesha make you an all-out winner in the realm of management, work, efficiency, handling promotions. Your personality will develop many new facets and the good results will naturally follow. This will be the trend for the month.

27 January: Moon's Last Quarter in Scorpio

Opposition to plans and projects may be likely, making it difficult but not impossible to fulfil your commitments. Alliances, even marriage, and publicity are, paradoxically, also important. Tolerance and foresight and, above all, getting your priorities right are essential.

4 February: New Moon in Aquarius

The Moon may also cause some tensions and ill health. Not major worries but nagging troubles. Attachments (as in the previous forecast) and journeys are the main areas of activity. Marriage and the home too, adds Ganesha.

12 February: Moon's First Quarter in Taurus

House, home, land and property are the focus this week. A home away from home may materialize. Investment plans, buying and selling, loans, machinery keep you busy in the money angle. Home and family, parents and in-laws will also make demands on your time and money.

19 February: Full Moon in Scorpio

A strange phenomenon is at work this week – you will be given to sudden likes and dislikes which will be quite pronounced in intensity. Some important decisions will need to be taken. An emotion-packed week as well. Don't let your feelings run away with you, warns Ganesha. Your decisions will affect the course of your life – and this is a monthly trend.

26 February: Moon's Last Quarter in Sagittarius

You open up this week to new ideas and social interaction. This glasnost is a new facet for Leos who have definite ideas and views. Group activities, visitors and relatives and some romance add a touch of spice to your life, and that's wonderful!

6 March: New Moon in Pisces

Loans, funds and religious ceremonies are favoured by the Moon. Moderation and care of your wealth will be more necessary than ever. Hasty pronouncements, rash actions, brash talk *must* be avoided or you may rue the day you made them!

14 March: Moon's First Quarter in Gemini

Your creativity and the romantic side of your nature are greatly enhanced this quarter. Travel and spiritualism will excite you. You will be far more relaxed, laid-back, and much more fun to be with. Your personality will flower and please those around you.

21 March: Full Moon in Libra

The Moon strongly and firmly favours money as a trend. Buying/selling, doing the groundwork for new projects and future gains will widen to encompass both work and recreation. If handled wisely, you'll reach a pinnacle from which there will be no looking back.

28 March: Moon's Last Quarter in Capricorn

The trend of last week will continue this week too; in fact, gain momentum where new projects are concerned. You will be receptive to new skills, ideas, knowledge, research and also equally so to spirituality. Creative endeavours, love and children's affairs will all benefit.

5 April: New Moon in Aries

You will gain much in balance and maturity, becoming more open-minded. You will look for ways of bettering your financial status, but will do so quietly and quite unobtrusively. You may also have to deal with hindrance/disappointment and conflicts. Please, reread the first sentence of this quarter's forecast.

12 April: Moon's First Quarter in Cancer

This phase of the Moon ushers in a different mindset. From this quarter, your perspectives will have changed, and so also your prospects. A tremendous step forward – and as I said earlier, no looking back. The main focus will be on partnerships and alliances, a kind of power brokerage, and trying to get ahead.

18 April: Full Moon in Libra

The social sense starts looking up. Entertaining, socializing, travelling, communicating on a personal level, especially with neighbours, kin and friends are all greatly enhanced. Community projects, the larger view, and love too. A monthly trend.

26 April: Moon's Last Quarter in Aquarius

Employment is the primary focus – no matter whether you're employed or not. Hard work and matching rewards sum it up. Doing your duty, giving happiness to others, personal satisfaction, great creativity for artists, writers, musician – these are what Ganesha bestows, along with general welfare and enthusiasm.

4 May: New Moon in Taurus

The theme of last week will continue into this one too. Personal progress, ambition, rewards and recognition figure prominently this week. Also, the home front, parents and in-laws, in fact, the whole and entire scope of activities that make up one's life.

12 May: Moon's First Quarter in Leo

A tremendous, in fact mind-boggling and gravity-defying leap upward in work and emoluments, in status and prestige, in confidence and being successful. Some or all of the above will be strongly influenced. A true pole vault, soaring up. Romance and love too.

18 May: Full Moon in Scorpio

Loans, funds, finances have to be carefully orchestrated for future gains. Those employed also do much better. A specially favourable time for shifting and moving office/home/city. Even immigration may be likely, if you are so inclined. This is also the ideal time for relationships of all kinds.

26 May: Moon's Last Quarter in Pisces

News, views, interfacing with people, fresh new contacts and channels of communication, all get energized. It's also time to receive, accept, give and interact. Even coming together or separating permanently are possible. Ganesha suggests that one word for all this could well be *people*.

3 June: New Moon in Gemini

Finances and loans, new avenues of earning, and income generation are once again at the forefront. You will definitely be building up financial resources and may strike good bargains, favourable deals and partnerships. A lot of all this may materialize over and through wining, dining, hospitality.

10 June: Moon's First Quarter in Virgo

A very personal, intimate phase and month for you starts now. Family, parents, in-laws, children, relatives, even a gathering of the clan keep you busy. Retracing your family tree, perhaps even religious rites are foretold. A development that is certainly different, if not unique, for you.

17 June: Full Moon in Sagittarius

Work, work and again work, says Ganesha, is the theme of this quarter of the Moon. Not just hard and unrelenting, but a total preoccupation. You won't be going it alone, though. Secret help, counselling and advice may materialize. Health care will be imperative. Burnouts went out of fashion!

25 June: Moon's Last Quarter in Aries

Travel on the one hand and, in a total about-turn, home and hearth and family on the other are the main focus this quarter. Alternating bouts of intense activity and relaxation are likely. Office and home may well come together, perhaps literally under the same roof. A great surge of creativity and artistry – for those so inclined – is foretold by Ganesha.

2 July: New Moon in Cancer

Fun and games continue to be fused with hard work, as they were right through June. You will certainly get a lot out of life that way. Social life and entertainment and yet sustained hard work, both continue to keep you not only gainfully occupied but happy with it.

9 July: Moon's First Quarter in Libra

Partnerships, be they marital, business or professional, are your major concern. I've said 'major', not 'only'. Personal issues like children, your own health and mental/spiritual well-being, your total persona will all occupy your mind. You realize there can be no peace, no contentment, no joy without adjustments and compromises.

18 July: Full Moon in Capricorn

The inner world, the inner self, will compete neck to neck with trade, finance, hypothecation, contracts and collateral. Loans will materialize. Taxes, rentals and outflow of money too. You will be in for a fair amount of entertaining, partying, eating out, both for business and pleasure. A slightly tricky month, with interwoven trends.

25 July: Moon's Last Quarter in Taurus

This quarter will strongly focus a yearning for peace and tranquillity, some other personal issues, along with last week's trend of loans/funds and such. House, home or office space may also require your attention, and possibly entail some expenditure. Vehicles and means of transport too, probably right through till the end of August, advises Ganesha.

1 August: New Moon in Leo

Your expenses will be tremendous once again. The home/office will also definitely figure in the scheme of things. Property, land, vehicles will be important, and of course, strictly personal matters. You will find yourself pulled in different directions. But all this will teach you to be confident, focused and strong.

7 August: Moon's First Quarter in Scorpio

This quarter bestows both money and good times. I have frequently noticed – and pointed out, dear readers – intensified activity in the birth month of a sign. Along with money and fun will be love and romance, quite possibly a marriage or an engagement, and some praise and appreciation.

15 August: Full Moon in Aquarius

This quarter steps up all of the above, and also work, dependents, subordinates, pets, children, colleagues. Both work and enjoyment in great quantity. Also, a kind of social awakening to larger issues – of service, welfare and charity.

30 August: New Moon in Virgo

Alliances, collaborations, partnerships, marriage, engagement – one word, *ties* – sums it all up. Lots of money inflow, but unfortunately also expenses and outflow. Money is money only if spent, says Ganesha. Otherwise, perhaps, the root of all evil.

6 September: Moon's First Quarter in Sagittarius

The impact of this phase of the Moon will carry over right through September. Love, yet strangely enough, a loss of face or reputation or just plain bad-mouthing, a fair amount of travel, perhaps even a major move to a different place, are all foretold. Keep your options open, advises Ganesha, and don't burn all your bridges – even those of communication.

14 September: Full Moon in Pisces

This phase of the Moon carries over, as already stated. Additionally, your work, efforts, endeavours and planning now bear fruit. Recognition, rewards, promotion may materialize. Quite a high-powered week, or at least one that will lead up to it.

22 September: Moon's Last Quarter in Gemini

A wish-fulfilment, some joy. And a degree of satisfaction. It may be just in a job well done or the success of a venture or enterprise. Culmination and fruition can be considered a trend for the month, ushered in now. You will be sure of succeeding, of going places, of making the right moves. I hope you'll do so as important papers, deeds and documents are also highlighted.

28 September: New Moon in Libra

This time around, your angle of finances, family and travel all gain momentum and force. You feel that everything is finally moving, you're getting somewhere. Advancement of learning, acquiring a new skill or qualification are also likely.

5 October: Moon's First Quarter in Capricorn

The full Moon strongly emphasizes health issues and family matters. In fact, it's a kind of ropeway or bridge linking the two themes. The emphasis, however, is firmly and strongly on financial and business matters – which will become even stronger in the coming month.

13 October: Full Moon in Aries

This quarter is for contacts, the home and family, and matters pertaining to marriage and divorce. Relationships are, therefore, all-important this week. Travel and trips with a stopover are likely. Personality development and interacting in human relationships will be strongly focused.

21 October: Moon's Last Quarter in Cancer

Drive, determination, enterprise are what Ganesha tells you to capitalize in this quarter. You may even need to be clever to the point of craftiness, devious and secretive. In short, what I'd like to advise you is to keep your own counsel, play your cards close to the chest. The astrological reason is that others may be watching for one false move, a slip on your part. A strange week, but that's what Ganesha reveals to me – and it's a trend.

28 October: New Moon in Scorpio

You have to make all-out efforts to win. You may need to snatch or wrest the advantage, the way Pete Sampras did at Wimbledon last year, to equal Borg's record. There's a lot at stake (plenty of money too) and the effort has to be made. The financial umbrella extends to cover this week too. But it also includes a strong focus on the entire range of human relationships – parents, children, family, kin, even society and the world.

4 November: Moon's First Quarter in Aquarius

An easing of tensions and complications will be noticeably perceived. Not only will you feel better but you will also think, act and be better. Better health too, positive thinking, good deal of love and laughter, fun and games, social interaction, quiet confidence and successful meetings – lots of goodies from Ganesha.

12 November: Full Moon in Taurus

You have the deadly power and resourcefulness that many a don of the Cosa Nostra (a US criminal organization related to the mafia) would envy. What else can I say except that you get where you want to be, regardless of what sphere you act in. House, home, property, business and spirituality are strongly emphasized.

18 November: Moon's Last Quarter in Leo

The Moon finds you more than holding your own on the fast track. Enthusiasm can have far-reaching effects, of course, but there is the hidden danger of overreacting and of overextending yourself. A new beginning is possible, in spheres of creativity, perhaps even starting a family. Some kind of wish-fulfilment too.

26 November: New Moon in Sagittarius

A strong leaning towards spiritualism is coupled with a strange restlessness that may manifest itself in frenetic activity and work. You may even take on extra work. You will be outstanding in research and invention, if so inclined, especially in the fields of electronics, automation, computer-aided design, graphics, and multimedia. Journeys are likely too.

4 December: Moon's First Quarter in Pisces

You look inwards, during this phase of the Moon, into the kingdom of the self and the human spirit. At the same time, you will be extroverted, communicative, sociable. The focus of all this is definitely going to be all activities pertaining to house and home.

14 December: Full Moon in Gemini

A very fine week, in terms of both work and play, but also for getting a kind of guidance just when you need it. Better health, greater productivity, ties and bonds (both with your spouse and children) promote a sense of well-being and a feeling of confidence in yourself, and the future.

19 December: Moon's Last Quarter in Virgo

The Moon will favour secret connections, even conspiracies, and string-pulling, but the net result will be achievement of objectives, getting to the top. Journeys, education, children's affairs, sports and hobbies are important. I would personally interpret them to mean different ways of extending yourself, of reaching out. And that's a fine trend indeed, for the last month of the year.

26 December: New Moon in Capricorn

Research, publicity, ceremony, journeys, religious rites, foreign connections, artistic and inspirational activities are all favoured. You will be powerful and eloquent in speech. A strange gift of prophecy/second sight might be yours too.

VIRGO

23 August–22 September

Virgos are the ants of the zodiac. They work hard, carry the heavy load of responsibility and most certainly do their duty by one and all.

Virgo is the sign of the Critic, the Secretary. Virgos are exact and methodical, analytical, sincere, methodical, psychic, rigid, prudish, fault-finding, miserly.

I am repeating for the last time that Narendra Modi was born to cleanse and purify the rivers and minimize pollution. In fact, Virgo and Scorpio, by Western astrology, are meant for this. At eighty-seven I have no reason to flatter anybody. The sign Virgo is derived from the Virgin. The Virgin is clean, pure and sacred. All these three attributes go with your sign. That is a very big deal.

This time I am following it up with your digital plan for the year so that you can focus very well on it in advance and thus prepare yourself by making a blueprint of the entire year. The digital plan for the year will be short and to the point, and though it is a general reading, it should apply quite a bit to you.

BLUEPRINT FOR THE ENTIRE YEAR

January: Love, romance, the luck of the draw, creativity and children, family and fun and fortune.

February: Health, employment, pets, servants and colleagues, and a few problems connected with these.

March: Love and hate, marriage and making merry, but paradoxically, in a few cases, separation and legal cases. That is why life is so complex, uncertain, full of contradiction and surprises.

April: Funds, loans, capital formation, shopping.

May: Inspiration, journey, name and fame, good luck.

June: Power, prestige, parents, profession, awards and rewards, money, home, house and office.

July: You will be off and away to a flying start, says Ganesha, and as you know, well begun is half done.

August and September: Expenses but also progress, therefore mixed results.

October: Excellent for finances, family affairs and earned income.

November: Reaching out to people through all media of transport and communication.

December: Important for peace, buying, selling, renovation, decorating; a home away from home.

This time we are giving you extra information about your own sign. It will be useful forever and ever. Ganesha asks for a look at the decans. Ganesha says each Sun sign occupies 30 degrees of the zodiac. Further divided into three, they form decans of 10 degrees each, a study of which can yield further insights into your personality.

If your birthday is between 23 August and 2 September (the first decan), Mercury governs all your activities and occupations, making for perfection and precision in work, with sharp intelligence and manual dexterity. You are more friendly and sociable than any other Virgos, and do well at teaching, public speaking, interacting with people.

If you were born between 3 and 13 September (the second decan) you are ruled by Saturn. You are very practical and stable but can also be stubborn and wilful, carrying your determination too far. You are very hard-working and conscientious, doing extremely well after the age of thirty. You may have weak bowels and a tendency to melancholia. Money means a lot to you, so you may even be a miserly penny-pincher.

If you were born between 14 and 22 September (the third decan) your ruler is Venus, and you gain both pleasure and profit from artistic activities. You have both refinement and a sense of fine discrimination and are much more versatile than other Virgos. Don't become a perfectionist or expect too much from your companion/life partner.

Ganesha says, yes, your heart will rest in love, friendship, and happiness in your home. Why? From 9 March 2018 to 2 December 2019, Jupiter will be in your fourth angle of:

- Buying/selling of house/office/shop/godown/warehouse;
- Renovating/decorating/altering/adding a room or so;
- Better health and more cheerful ambience/surroundings;
- Taking an active interest in all your surroundings, as we Indians say, but that does not mean that you have

to be nosy and interfere in the affairs of others, though most of us also do this in India! It is the done thing! Journeys, a home away from home, are also probable;

- Community, possessions and family will definitely, oh, very definitely, take your time and talent. Ceremony and social events give delight;

- Relations with parents, elders, in-laws, family members, making your will and codicil if you are an elderly person, for the simple reason that the timing is right, and therefore, you will be able to do justice to all your dear ones.

SPECIALLY FOR YOU

Saturn will be in your fifth angle from 21 December 2017 to 23 March 2020 but this will very specially apply to you in 2019.

> May I be like the Sun in seeing; like Fire in brilliance; like Wind in power; like Soma in fragrance; like Lord Brihaspati in intellect; like the Ashvins in beauty; like Indra–Agni in strength.
>
> —Sama Veda

Let me now go into some detail:

- Saturn highlights your creative impulses, your romantic and artistic needs, your sense of untrammelled, unhampered, total freedom. Children and grandchildren give joy. Entertainment and amusements will be your birthright. You could be moving with the younger crowd, maybe the jet set, maybe students, maybe toddlers and kids. Or you may be writing about them, painting them, and in any case having much to do with them. That's the bottom line. Do not become involved, with the problems of others, not now, please. Also, you could lose the affection of others, yes, that's the flip side. And, there is scope for a suitable avocation.

- Secret knowledge, journeys, an initiation, travel, spiritual and uplifting experiences, which just dissolve the dark aspects and recesses of life and sorrow, will be yours. Yes, you will overhaul your entire life and make it grand and complete. It is like a vacuuming from within and after that, how can there be any looking back?

- 'Perpetual mobile' is the Latin for something in perpetual motion, and that means a grand reaching out to people and places, an ability to move, manoeuvre and manipulate, becoming precise and penetrative in letters, calls, communication, contracts, coexistence. You will interface with people powerfully and pleasurably, and that, by itself, is rare and noteworthy.

- So strong is this movement and this reaching out that it could result in a house or office or shop or industry move, immigration, contacts with foreign lands, neighbours, society in general. You could almost step out of your body and do some astral flying. I am not kidding! Move it, that is the glorious and clear message from Saturn.

Your happiness quota will be 88 per cent.

SPECIAL BONUS

Ganesha says my devotee Bejan is introducing the topic of special bonus for all of you. This special bonus is placed upon Mercury. Mercury represents journeys, ties, trips, mobility, migrations and all types of communication. Venus represents love, beauty, the fine arts, polish and finesse, diplomacy and persuasion, joy and a big hurrah to life itself. Therefore, I am combining both communication and joy for you and giving you the necessary dates.

Mercury is your main planet. Therefore, I am taking it up first. Mercury will help you from 5 to 23 January (fun and frolic), 10 February to 16 April (love and romance), 7 to 20 May (fulfilment of your heart's desire), 5 to 26 June (friends help you), 19 July to 10 August (be social and friendly), 29 August to 13 September (perhaps the best period of all), 29 December to 16 January 2020 (happy times are here again).

Venus will help you from 4 February to 1 March (romance and finance), 27 March to 20 April (creativity and journey), 15 May to 8 June (children, hobbies, entertainment, love), 4 to 27 July (party time, happiness in home), 21 August to 14 September (your dreams may come true), 9 October to 1 November (travel and joy), 26 November to 19 December (happiness will be your great friend).

However, I must be true to my own self. This is only a solar-scopic reading based mainly upon the position of Sun. Therefore, it may not be as accurate as a personal horoscope. At the same time, I have put in a lot of effort, imagination and inspiration as well as intuition into it. In other words, I have tried my best. The results are up to our maker. Yes, sometimes miracles do happen. Therefore, I end on a positive note, though I know my limitations very well.

KEY TO WEALTH BY NASTUR DARUWALLA

Virgo are great in building relationships. They are by nature honest and hard-working. The mantra for them is: '*Om hrim shrim klim mahalakshmyai namah*'. Help the needy giving school fees.

WEEKLY FORECAST BY PHASES OF THE MOON

6 January: New Moon in Capricorn

Get-togethers, unions and reunions, any form of group activity, are favoured. You can do justice to any position of responsibility/trust offered to you. You may, however, have to assert yourself strongly for your rights and for your love.

14 January: Moon's First Quarter in Aries

The focus is on finance and loans, funds, retirement benefits, gratuity, securities, joint finances and bonus. You will make a go of things, and feel you're getting somewhere in life, even getting your just rewards.

21 January: Full Moon in Leo

Release from tension, relief from worry, freedom from restraints and opposition, romance, journeys – there is all that and more to make you feel on top of the world. The 'and more' of the last sentence means money, wise investments, better prospects. A really fine trend for the beginning of the year.

27 January: Moon's Last Quarter in Scorpio

You will find yourself acceptable to one and all. Popularity will be at an all-time high. You will have plenty of practical knowledge and shrewdness, class, polish, intelligence and wit. Romance will therefore be thrilling, and wonderful career moves and money matters all handled with much insight and wisdom. Luxuries, comforts, a sense of well-being come as gifts from Ganesha.

4 February: New Moon in Aquarius

Pending assignments and work will be completed, new ones started. Deeds and documents, contracts and communications will keep you busy – also, travel, visitors, relatives. Some kind of goal realization, if not wish-fulfilment, is likely. The strongest element will be the personal one, though.

12 February: Moon's First Quarter in Taurus

You will feel that your batteries are recharged and you are ready to attack life with renewed gusto and enthusiasm. You will be generous with favours, advice, hospitality. You will have the capacity to do the right thing at the right time. Very strangely indeed though, you will feel a tremendous feeling of being alone. Health may also cause you a few twinges to anxiety.

19 February: Full Moon in Scorpio

Finances and clandestine and secret activities will be at the forefront. You will be introspective, meditative, searching for a meaning to life and your place in the universe. You will, however, be sociable and gay at the same time – performing on two levels, as it were. Strange and exciting!

26 February: Moon's Last Quarter in Sagittarius

Your introversion continues, but worldly considerations will take up a lot of your time, energy and attention. Also, money, pets, your own performance and slight ill health, and both big and small projects for the future. All this requires an extroverted approach. That means you are living on different planes at the same time. Exciting!

6 March: New Moon in Pisces

A week of tremendous swings, of being up one minute and down the next, a real roller-coaster ride. There may be problems in both your personal and professional relationships. There will be good news and prosperity on one hand and delays and disappointments on the other.

14 March: Moon's First Quarter in Gemini

Another quarter of mixed influences and results. This is true of most of this month for Virgoans. Another quarter of mixed influences and results. This is true of most of this month for Virgoans. Journeys with a stopover, marriage (or problems therein), legal matters and collaborations, mishaps and creative pursuits – all make demands on you.

21 March: Full Moon in Libra

This phase of the Moon is a period of healing and coming to terms with yourself; of teaching yourself the karmic lesson. A period of inner rebirth. This is not to be dismissed as an exaggeration, dear Virgoans. It is a clear message from Ganesha, my Lord, Master, Mentor, Perceptor.

28 March: Moon's Last Quarter in Capricorn

You could run into problems this week, either personal or at work. Or even both. You may need to deal with separation, even death. Accidents and injuries, however small, cannot be ruled out either. Riots, losses, natural calamities, danger from the elements can affect you. Take heart, though, this may be your last tunnel before emerging into the sunlight, and the last week equips you to cope.

5 April: New Moon in Aries

After the healing process that was initiated in the earlier quarter, you now feel that you are in a position to move far up the evolutionary ladder. Health, mind, heart and spirit all feel richer and fuller. All your latent qualities will not only emerge but shine through.

12 April: Moon's First Quarter in Cancer

A phase of attending to strictly practical matters. You will be level-headed, pragmatic, efficient, and derive much benefit from it. The 'real world' as they like to call it, reclaims you and demands management skills, and organization techniques, but it may yield perks and promotions.

18 April: Full Moon in Libra

This is a phase when you will be taking chances. You will certainly have an easier time now. A slight cash crunch may be experienced, but the flow of money will never really stop. On that score you will have no cause to worry, in fact, you will enjoy life to the hilt, 'drink life to the lees'.

26 April: Moon's Last Quarter in Aquarius

A good time for people in the artistic and creative world – artists, painters, even chefs, connoisseurs, art critics and dealers, all those who are experts in the art of good living. Family welfare and financial transactions are also important this quarter.

4 May: New Moon in Taurus

The main thrust and focus will be strongly on employment – new avenues, changes, a job if you don't have one already. It can be summed up as a phase of hard work and matching rewards. A phase which favours advancement and job satisfaction, also the satisfaction of having done your duty. Travel and visitors, foreign connections too.

12 May: Moon's First Quarter in Leo

You will be introspective, maybe even restless and yet full of new ideas and innovations. There can be ingenious schemes and concrete opportunities to make money. Far off, distant places could have a connection, so it might be export–import, overseas projects that interest you.

18 May: Full Moon in Scorpio

This phase of the Moon will now bring all your activities of the last two quarters into sharp focus. You will be taking up new assignments, making new commitments and making a lot of headway in the field of contacts and communications.

26 May: Moon's Last Quarter in Pisces

The right time to cash in not only on the contacts and communication I spoke of just above but also for study, experiments (and their practical applications), research, new theories and ideas. This may entail not only travel but also the readiness to take chances.

3 June: New Moon in Gemini

You will work like a navigator this coming month, and specially from this quarter onwards. Progress, free-wheeling and fun are all favoured now. You will have to push, and this month can yield results like fatter pay packets, better perks and such.

10 June: Moon's First Quarter in Virgo

Parents, in-laws, boss/superiors will tend to slave-drive you. But the rewards will be there too. A favourable time, says Ganesha, for taxes, rents, renovation, decoration or perhaps even a new job or house. Also, equally important will be relatives, even a gathering of kin. Some rites or ceremonies are likely too.

17 June: Full Moon in Sagittarius

This phase of the Moon accelerates all the trends described. All this will come to a head around July or August. Changes are likely at home/office/workplace. Anything to do with progress and achievement (perhaps even the lack of it) is highlighted. Professional rewards, status and prestige too. Your popularity will also register this.

25 June: Moon's Last Quarter in Aries

In most respects, a definite continuation of the previous quarter. Honour and status will be major concerns. Profession/business, getting on in life, will be the focus of all you say, do, plan and implement.

2 July: New Moon in Cancer

You clear away all the doubts, despair, even encumbrances to progress. However, all this will require almost superhuman effort but you will summon up both the energy and the motivation to do it. The good part will be that you will love and be loved, despite your ambitious thrusting ahead.

9 July: Moon's First Quarter in Libra

There may be some clashes with people of a different temperament. There will be heavy emotional responsibilities, linked to a chance of promotion, a further step up the ladder. You must learn to show people you care. It pays – for yourself and for them too.

18 July: Full Moon in Capricorn

You will be an ardent lover, also a lover of your country. Fantasies and sentiments will fascinate you, grip you. You will also acquire a good deal of tact and glamour which will be of great help not only in inner personal relationships but in your work too.

25 July: Moon's Last Quarter in Taurus

A time of swashbuckling adventure, travel, piracy on the high seas of business. You will climb the ladder of success and will be totally work-oriented. Health care will prove to be an absolute must – in fact, right through for the next four weeks, at least till the end of August.

1 August: New Moon in Leo

You will display great sensitivity and imagination and charm in all you say and do. Opportunities for business can thus be encashed, especially with regard to trade/brokerage/buying/selling. Also important will be renovation/decoration of the home/office. You will do everything in style, including entertaining.

7 August: Moon's First Quarter in Scorpio

An ideal time to be outgoing and forward, social and amiable. You will please not only yourself but others also at the same time. You will get your work done, and yet keep others happy. The fine arts – music, painting, sculpture, drama – will be particularly rewarding. Also, any work connected with the arts will thrive and prosper around this time.

15 August: Full Moon in Aquarius

This phase of the Moon connects with work, projects, colleagues and subordinates, and pets and animals. This is definitely the trend for the month – a very high-profile time for you. Social service and commitment will be strong within you, but though you work like a beaver, you will also have a royal time enjoying yourself.

30 August: New Moon in Virgo

Your emotions will be churned up in many ways. You will be having to perform miracles, not only to please your boss, but live up to your own image to yourself. An engagement, wedding or love affair is also more than probable – if so inclined. Journeys and contacts are also favoured. See what I meant about being churned up? This quarter and the one to come will have you spinning like a top.

6 September: Moon's First Quarter in Sagittarius

To add to the frenzied activity of the last quarter will come collaborations, export–import, trips abroad, studies in a foreign land, psychic phenomena, surgery, history and research. The possibilities, as they say, are endless. You will have a tough time choosing.

14 September: Full Moon in Pisces

Buying, selling, brokerage will be the main areas for financial advancement, including furniture, and vehicles, what are called 'white goods' and consumer durables. This may well be part of the theme of renovation/decoration/upgradation of the house/home/office. You may even be attending sales and auctions.

22 September: Moon's Last Quarter in Gemini

This phase of the Moon is for ties and bonding, not just with those 'near and dear' to you, as we say in Indian English, but with the world at large. What a trend! People from distant places and you will experience mutual joy in meeting. Television and the film industry may even draw you in – you might be in the news too!

28 September: New Moon in Libra

You will put in that extra bit of effort, that last push towards fame, recognition. You will gain acclaim and rewards not only at home and at work but possibly even abroad. A feeling of finally getting your deserved and just rewards will spur you on. Glory to Ganesha for making it happen.

5 October: Moon's First Quarter in Capricorn

This is a phase of transition to October, astrologically. You will experience rare insights and self-improvements, an overwhelming upsurge of energy and achievement, and a move down to earth, changes or moves or shifts at home and/or work. The emphasis will be strongly on finance and health, now and practically till the end of the year.

13 October: Full Moon in Aries

You are at your scintillating best, charming all around you and making friends as you go. At the same time care of work/home premises, of children is a major concern, along with gadgets and machines to enhance and improve your lifestyle.

21 October: Moon's Last Quarter in Cancer

The new Moon stresses, underlines, reiterates the trend of last week, develops it to true strength and even extends it further. In addition to all that's happened over the last few weeks will *also* be a passion, a joy in love, sex, and intimacy that is rarely found. A passionate love affair may obviously emerge.

28 October: New Moon in Scorpio

You will be romantic, both attractive and attracted to the opposite sex, in a carryover from the last quarter. Marriage, affairs, liaisons are once again predicted. Along with romance will come career advancement, material and monetary gain, also prestige and power, you are riding the crest of a wave and what a wave!

4 November: Moon's First Quarter in Aquarius

Luxury, comfort, opulence, the heights of style, living, food, society – you sample them all. It is one of the finest months that anyone can hope to have, as you look back on November. Family, entertainment and amusement are also favoured by Ganesha who laughs as he spills his goodies out.

12 November: Full Moon in Taurus

You will score heavily in many directions, once again. Dealings with younger brothers and sisters, neighbours are favoured. Also intellectual attainments, tests and interviews, transport and, of course, the three Cs of communications, contracts and correspondence. You will score heavily in all these directions, so once again, what a week!

18 November: Moon's Last Quarter in Leo

You are launched into the big-time league, provided you can talk, converse, communicate, use your contacts. All your activities receive a terrific boost – and you may have to do a balancing act to manage them all. Journey, ceremonies and rites, marriage and/or engagement and publicity are part of the action.

26 November: New Moon in Sagittarius

You have to gird your loins, not to go into battle but to prepare for consistent hard work, which will pay remarkable dividends, both in financial and other ways. You will earn both power and appreciation and also polish your managerial skills, boost your confidence. You will be both giving and carrying out orders.

4 December: Moon's First Quarter in Pisces

The full Moon has two diametrically opposite influences. Mind-boggling involvement with contacts, communication, correspondence on the one hand and a search for the kingdom of the soul and spirit, on the other. You will give priority, however, to everything concerning house/home and property – from buying and selling to renovation and doing it up.

14 December: Full Moon in Gemini

Divine and timely guidance seems to be yours, as you perform in several spheres – new enterprises, expansion of existing assets, neighbours and children, higher learning and research, communication and travel, even the fine arts, not to mention love. It seems to me that you have Ganesha's supreme blessings!

19 December: Moon's Last Quarter in Virgo

Personal matters like marriage, children, meets, functions and conferences thrust you willy-nilly into a round of hectic activity. You not only cope but acquit yourself with style and grace. The new Moon also gives you just the right power thrust in not only to work and recreation, but also in foreign connections.

26 December: New Moon in Capricorn

You are both happy and satisfied this week (the two don't always to go together, believe me!) The pressures on you will gradually decrease, and health problems will improve despite a quarter that is emotionally charged.

LIBRA

23 September–22 October

Libra is the sign of the Judge, the Ambassador. You Librans are diplomatic, charming, easy-going, balanced, indecisive, detached, fair-minded, alert, changeable, artful.

To me at least Amitabh Bachchan, Amzad Khan's character in *Sholay*, Lal Bahadur Shashtri, Mahatma Gandhi, and A.P.J Abdul Kalam, the former president of India, show the wide range which you people cover with a single gallop. You can be not only good but also great. That speaks for itself.

This time I am following it up with your digital plan for the year so that you can focus very well on it in advance and thus prepare yourself by making a blueprint for the entire year. The digital plan for the year will be short and to the point and though it is a general reading, it should apply quite a bit to you.

BLUEPRINT FOR THE ENTIRE YEAR

January: Home, house, family, shopping, renovation, decoration, alteration.

February: Trips, ceremonies, rites, name and fame, future plans.

March: Work, health, rewards, services, pets, projects, colleagues, promotion.

April: Love, marriage, lawsuits, relationships, travels, contracts, communication, enemies.

May: Joint finance, loans, funds, immigration, moving, shifting, capital formation, passion, sex.

June: Joy, publicity, travel, ceremonies, functions, parties, irritations, import and export, collaborations, contacts, contracts, happiness because of abundance.

July: Up and about, hard work, you will be ready to be on the go be it home or office.

August: Fun and games, friendship and fraternity, gains and gaiety.

September: Too many things happening all together, need for conservation of energy/vitality.

October: Progress, prosperity, getting things done.

November: Loans and funds, deals and transactions, buying/selling.

December: Assignments, communication, transport, ties, trips, relatives.

This time we are giving you extra information about your own sign. It will be useful forever and ever. Ganesha asks for a look at the decans. Ganesha says each zodiac signs occupies 30 degrees, which can be divided into three decans of 10 degrees each. Although all those born between 23 September and 22 October are Librans, the individual traits vary according to the decans.

If you were born between 23 September and 3 October (the first decan), your planet is Venus, and beauty and harmony are everything to you. You are the peace-maker par excellence, love is everything to you. You love to impress people and, given congenial surroundings, you do outstandingly as well at your work.

If you were born between 4 and 13 October (the second decan), the planetary rulers are Saturn and Uranus, making you an unpredictable but strong person, full of magnetism. You are a great champion of a new order, with striking powers, and the ability and insight that you possess take you very far.

If you were born between 14 and 22 October (the third decan), your ruling planet is Mercury, giving you tremendous energy and power of persuasion. A special flair for writing is seen, as also, teaching, journalism, advertising, publishing. However, learn to be more precise and firmer.

Barack Obama, the former president of America always said that all of us are 'interrelated'. This actually applies completely to you because Jupiter will be in your third angle from 9 March 2018 to 2 December 2019. Jupiter, the planet of plenty, prosperity and wisdom (a rare mix) whirls away in your third angle of achievements and approbation. The third angle signifies:

- Connections, contacts, reaching out to people and places;
- Journeys, ceremony, legal matters;
- Public relationships and competition, change of locale/surroundings;
- Collaborations and cooperation, wedding/engagement;

- Blowing hot and cold in relationships.

You will have to hurry your strokes, as they say, work fast and furious, complete deals and negotiations and settlements. It is not that you will have no chance to do so after December 2019. It is just that this period is perhaps your best bet.

The third angle represents the hands. Therefore, in cricket all the bowlers, and for women all the fashion designers and hair dressers, also come under the direct influence of Jupiter in the third angle. In Indian astrology, the following are also added:

- Gains to brother/sister/neighbours/cousin of self or of spouse on death or renunciation of *garhasthya* by self, and determination of point of time for such an event;
- Marriage with fiancée/widow/widower/divorced spouse of brother or sister or cousin, or his/her maintenance without marriage, or any kind of immoral or sexual relationship;
- Reward/award/decoration/grant of title/grant of land or any other immovable or moveable asset/grant of family pension or hereditary pension for acts of gallantry, heroic bravery, outstanding leadership in a battle or for spying and intelligence activity.

Saturn, the planet of limitations and perhaps also of suffering and sorrow, duties and responsibilities, spirituality and humanity, will be in your fourth angle from 21 December 2017 to 23 March 2020. Strictly speaking, the fourth angle stands for house, home, property, family, land, building, godown, warehouse, farming. In short anything to do with parents, in-laws, buildings, gardening and property. Here Saturn will blow hot and cold making things sometimes a bit difficult for you. Farming and mining, elders, guardians, the settling of outstanding matters, improving your environment, saving for the future will be some of the most important matters ever for you. If you are attached to elderly people, my strong advice for you is to look after them wisely and well.

In Indian astrology, the ancient acharyas call the fourth house the house of comfort or the house of parents, and the following subjects falls under its jurisdiction:

- Parent or step-parent, mutual relationship between parent and self;
- Of biological parent/step-parent/adopting parent, events and developments from it, subsequent impact on the individual and matters/problems of inheritance/succession therefrom and their settlement/solution or quarrels and litigation/violence, etc.;
- Acquisition by purchase, hiring, lease, contract, etc., of property or premises for own use or by other members of family or by partners, on own responsibility or liability and the question of relations between landlord and self-relating thereto;
- Breeding and maintenance of cattle, horses, donkeys, ponies, camels, goats, sheep, dogs, cats, elephants, hares, pet animals, birds like parrots, poultry farms, pigs, bulls, oxen, deer, antelopes, fish and fisheries, problems and disputes relating thereto, and using it as a side business or sole means of income.

Your happiness quota will be 84 per cent.

SPECIAL BONUS

Ganesha says, my devotee Bejan is introducing the topic of special bonus for all of you. This special bonus is placed upon Mercury. Mercury represents journeys, ties, trips, mobility, migrations and all types of communication. Venus represents love, beauty, the fine arts, polish and finesse, diplomacy and persuasion, joy and a big hurrah to life itself. Therefore, I am combining both communication and joy for you and giving you the necessary dates.

In your individual case your main planet is Venus – for pleasure, profit, diplomacy, the fine arts and your entire personality so to say. Venus turns favourable for you from 7 January to 3 February, 2 to 26 March (very specially), 21 April to 14 May (all types of connections), 9 June to 3 July (for hobbies, entertainment and children), 28 July to 20 August (for all social events), 15 September to 8 October (perhaps the best time to go right ahead in life), 20 to 31 December (for rounding up all events and happenings).

Mercury helps you from 24 January to 9 February, 17 April to 6 May, 21 May to 4 June (very specially), 27 June to 8 July (in every possible way) and most certainly 14 September to 2 October. I repeat Mercury stands for connections, contacts, trips and ties and also anything which is secret and hidden from the people at large. Mercury is your secret weapon. Use it wisely and well.

However, I must be true to my own self. This is only a solar-scopic reading based mainly upon the position of Sun. Therefore, it may not be as accurate as a personal horoscope. At the same time, I have put in a lot of effort, imagination and inspiration as well as intuition into it. In other words, I have tried my best. The results are up to our maker. Yes, sometimes miracles do happen. Therefore, I end on a positive note, though I know my limitations very well.

KEY TO WEALTH BY NASTUR DARUWALLA

Librans are charming, attractive and emotional people. Honesty and justice are major traits of their character. The mantra for them is: *'Om shrim shridevyai namah'*. Provide sponsorship to underprivileged children for all types.

WEEKLY FORECAST BY PHASES OF THE MOON

6 January: New Moon in Capricorn

You will be pushed into the limelight, the public eye. Matters related to house and home and travel will concern you and so also situations requiring your skills at organization and public relations. Love too will find a place in your heart. In any case, you can bank on the cooperation of friends.

14 January: Moon's First Quarter in Aries

You will not only work with a will but also see good things happening. Family-wise, it is a good time to draw close, heal wounds, bridge gaps and mend fences. Neighbours, brothers and sisters will all rally round you. It's also a comparatively easy time, financially.

21 January: Full Moon in Leo

You will consolidate your position and also make headway. A promotion or career advancement is likely as you are slated to make considerable headway. Messages and communication, news and views, contracts, travel and contacts will be unusually important.

27 January: Moon's Last Quarter in Scorpio

This is a good time to rake in the lolly, make money, as new avenues and vistas open up, in your job/profession. It is also a time for drawing dear ones together, restoring peace, and caring for children, maybe even the birth or conception of a child.

4 February: New Moon in Aquarius

This phase of the Moon augments your finances with the focus on loans, funds, joint accounts, securities, blue-chip investments. You will be entertaining and being entertained in right royal style – both for pleasure and for work-related matters.

12 February: Moon's First Quarter in Taurus

Friends, love, the good things of life are yours. Your work will get done smoothly as you are in a position to ask for favours. Success is when pleasure and profit are fused together – and that's the way Ganesha makes it happen for you this quarter.

19 February: Full Moon in Scorpio

This quarter of the Moon triggers off finances and communication with a mighty bang. You will have to move fast, take on-the-spot decisions and galvanize not only yourself but others into tremendous activity. All kinds of creativity, even if it's just creating new programmes on computers or cookery, is favoured.

26 February: Moon's Last Quarter in Sagittarius

Ganesha's command for you is an easy one to execute – socialize and circulate. This is the ideal time for that! While you enjoy yourself, advancement of career and prospects or a promotion are not ruled out either. A fine period.

6 March: New Moon in Pisces

A little health care is a must. You will have to relax, recharge your batteries, even as you continue to be ambitious and hard-working. You will also need to exercise some of your Libran tact and balance to sort out family problems.

14 March: Moon's First Quarter in Gemini

An ideal time for communication because at this time Mercury conjuncts Venus, your ruling planet. Calls, letters, advertising – the many and varied forms of communication – are all favoured this quarter, so also wining and dining and a good bit of socializing. This can be for both business and pleasure.

21 March: Full Moon in Libra

This phase of the Moon is for trips and ties. However, Ganesha warns you that accidents and mishaps are not ruled out. Air travel may not be your best choice at this juncture. It may prove to be hassle-prone and costly. It is definitely a time to be practical, and try and get value for money, even in ways other than financial.

28 March: Moon's Last Quarter in Capricorn

Though this is definitely a month for contacts and communication, paradoxically enough, it will be difficult to know where to stop. This will prove to be imperative, otherwise the desired results may not come through. A time to use your sense of balance. Some danger of ill health is possible.

5 April: New Moon in Aries

You may sense distancing and unease, fears and anxieties and feel more than a little threatened. The focus is on partnerships at all levels, even on fighting for position and power. It's *not* a time to be too pushy and aggressive. In any case, that's not your way, Librans, warns Ganesha.

12 April: Moon's First Quarter in Cancer

Poor health, separations and misunderstandings may have to be coped with, or perhaps legal issues. You may have to watch your step. Motives may be misunderstood, wrong signals sent out or received. Ganesha advises you to play it cool, play it safe – both professionally and personally.

18 April: Full Moon in Libra

You will be enormously productive and creative. Be careful not to overplay your hand, and you will succeed in all that you set out to achieve. Promotions and perks are likely. Home conditions and also your financial credit and bank balance need to be checked out.

26 April: Moon's Last Quarter in Aquarius

In practically all respects this phase is a carryover from the last quarter. The monthly trend of enormous productivity, work, yet a faint depression, a sense of physical and mental malaise will grip you. At the same time, you are wise and sage in the advice you give, displaying both inner maturity and deep understanding.

4 May: New Moon in Taurus

All the activities of May come together now – travel, contacts, interviews and dealing with problems too. The astrological focus, says Ganesha, is *people*, meeting them, reconciling with them, involving them in your many and various activities. Poor health will require the necessary safeguards.

12 May: Moon's First Quarter in Leo

The pleasure and profit principle will be firmly underscored. You will display fine taste, a love for the good things of life. All the comforts and luxuries that make lifestyle beautiful may come into focus – music, interior decoration, gadgets and furniture, even jewellery, tremendous socializing too.

18 May: Full Moon in Scorpio

A truly grand time for you sets in with this phase of the Moon. This luck will carry on right till mid-June. A wish-fulfilment is predicted by Ganesha, who says that this is when you get what you want. You may have put your troubles behind you now.

26 May: Moon's Last Quarter in Pisces

Your personal and career graph soars, in continuation of the trend of the last quarter. Happy times for those in sports, travel and tourism, the hospitality industry. Also, people with artistic talent may find themselves making a career of it, and doing well.

3 June: New Moon in Gemini

All the various media of entertainment and information (films and television, for example), will be used to your distinct advantage. You may find yourself in the spotlight. Romance and passion may also come your way. This month is really a very high point.

10 June: Moon's First Quarter in Virgo

This quarter seems to swing you into restlessness and introspection once again. Foreign connections, distant places, multinational and religious organizations, foreigners and research give your activities a global slant. Travel is also probable, and also love and/or liaisons in some form. There are, strangely, paradoxical forces pulling you.

17 June: Full Moon in Sagittarius

There will be plenty of movement, of a different kind right through. This is a monthly trend. A house move, a home away from home, the foreign slant, even import and export will be in strong focus. Also, your relationships with boss/colleagues, subordinates, even pets and dependents. The last two could pose some problems. You will develop a strong potential now for your intuition and extrasensory perception.

25 June: Moon's Last Quarter in Aries

The next two quarters will have only one, all important, overriding theme – *work*. You will look within yourself for that little bit extra, both physically and, more importantly, spiritually, and will come up trumps. You will be thinking, moving and working superbly.

2 July: New Moon in Cancer

Once again, work – but you will be able to finish pending assignments and you will have the added boost from the confidence that you can deliver the goods and cope with this competitive world we live in. Some secret help and guidance is also possible. *Health precautions are a must*, thunders Ganesha.

9 July: Moon's First Quarter in Libra

One of the busiest periods of your life is about to start and will carry on well into August. You may find yourself a beast of burden emotionally, psychologically and at work, but you not only cope but also manage to enjoy yourself, get more out of life at the same time. You will accomplish much.

18 July: Full Moon in Capricorn

Right up to August, starting now in this phase of the Moon, you will find that loans, finance, monetary transactions and buying and selling hold the key to a better future. You will receive help and guidance, not only from friends and well-wishers but from a higher, divine source. Ganesha chuckles. That's what really counts after all, because, as Shakespeare has so rightly said, 'The grace of God is gear enough' (gear here meaning wealth).

25 July: Moon's Last Quarter in Taurus

This should be a happy and memorable phase for you. Much will happen in terms of hard work, inspirational moves, contacts and correspondence. Right through, the heat will be just sufficiently on to make you aware of some rivalry. You will have to make allowances, give some leeway as people will not see eye to eye. They never do!

1 August: New Moon in Leo

Some confusion will have persisted in personal relationships. It will be a continuous struggle in many ways. Don't give in to rancour and regret, and you'll make things easier for yourself. A phase which is a medley of work and domestic affairs, attraction and repulsion, of travel and longing to stay home.

7 August: Moon's First Quarter in Scorpio

This quarter of the Moon makes you expressive and vocal to the point of being vociferous. Love and partnerships will be strongly favoured. It will be a terrific trio of money, romance and children, along with attachments at several levels that keep you involved and give a degree of satisfaction.

15 August: Full Moon in Aquarius

This phase of the Moon favours the money angle, in fact the entire gamut of financial activity – trading, joint finance, loans, funds, buying, selling, brokerage, interests and rentals. It's also a time to let go of yourself, even burn the candle at both ends, if so inclined, as you fuse work and play, chancing both your arm and your heart. May the luck of the draw favour you, says Ganesha.

30 August: New Moon in Virgo

This quarter could lead to an engagement, if not a wedding. The theme is alliances and collaborations; and even partnerships for business come within the ambit. I can best sum it up in one of my favourite phrases – money and honey!

6 September: Moon's First Quarter in Sagittarius

The impact of this quarter will naturally run over the entire month of September, and will provide a launching pad for one or all of journey and communication, a major move or shift, legal hassles and love, paradoxically enough. Around 22 September itself, a major stroke of luck might befall you.

14 September: Full Moon in Pisces

The ingress of Mercury into your sign from last week will continue giving you a different viewpoint, and perspective on life and living. The efforts and work of the previous few quarters could start showing results – promotions, rewards and recognition.

22 September: Moon's Last Quarter in Gemini

This is action time – an exceptionally productive phase vis-à-vis finance, journeys and travel, news and views, weddings and liaisons, ads and posters and one that may yield a rich harvest of fame and riches. Communications too will yield results – write letters, make calls, send faxes. A series of red-letter days till the end of the month, as you move into the month of your birth sign.

28 September: New Moon in Libra

I have frequently pointed out that the phase/month of any birth sign is unusually active and productive. So also, with you. From now through October, an unusual phase sets in – you will instinctively and naturally make all the right moves, the right decisions.

5 October: Moon's First Quarter in Capricorn

In many ways, a continuation of the last quarter, but with several add-ons. Visitors, strangers, foreigners will help you, trips will do good, affairs of the heart may blossom. Plans for the future are also favoured and the best thing for the last – divine guidance – is at hand, for some spiritual pursuits, and all your actions. Ganesha be praised.

13 October: Full Moon in Aries

The beginning of a phase that is exceptionally pleasurable *and* productive. This quarter is for alliances and contacts, home and family, and also for marriage, break-up or divorce. A journey with a stopover is likely. The theme could thus be: reaching out and relationships.

21 October: Moon's Last Quarter in Cancer

You have to push forward with all you have – a clear astrological message – this quarter of the Moon and for the whole month (I mean here the lunar month starting now). There's money in the game and you have to play to win, even if it means craft and guile, and behind the scenes activity, as others will not sit by watching idly.

28 October: New Moon in Scorpio

You are given a second, and even better, chance to make good. Personal affairs may give you a rough time. This is not the time to be sensitive and sentimental; you need to take off the kid gloves. There are strong chances that you will come out on top.

4 November: Moon's First Quarter in Aquarius

This quarter of the Moon will prove to be a real clincher, in terms of some or all the gains: romance, children, loved ones, group activities, entertainment and amusement. Equally so for social welfare, charitable concerns, attachments and collaborations. A memorable time for the young and not so young – a chance to make headway to find a place in the sun.

12 November: Full Moon in Taurus

You get an opportunity to complete pending work/projects and perhaps start a new venture. Both pleasure and profit are foretold. People from distant places and even foreigners may be of help. Trips and love affairs are both possible. Ganesha has a lot for you this quarter.

18 November: Moon's Last Quarter in Leo

Finances are strongly highlighted – the making and spending of money. I'm talking of big money here, and of buying/selling/shopping/bargaining/trading. There will also be satisfaction and joy from a family get-together. It's your personal charisma that makes all this happen.

26 November: New Moon in Sagittarius

In most ways, a continuation of the trend started in the previous quarter. It's a time to make a dream come true, or at least, get things moving your way. This quarter of November seems custom-designed for pulling together, working as a team, interfacing and interacting.

4 December: Moon's First Quarter in Pisces

Myriad possibilities lie before you but within the overall theme of collaborations there may be partnerships, liaisons, love affairs and proposals for marriage. It is time to take advantage of, and make the most of, once again, your personal charisma. The good things of life come your way easily.

14 December: Full Moon in Gemini

The magic will not peter out but continue in this quarter, too. Comforts and luxuries, love and money, new ventures and assignments, even an opportunity for the artistically talented. You might try to grow wings and fly.

19 December: Moon's Last Quarter in Virgo

A little practicality, some coming down to earth in the midst of your euphoria, so says Ganesha, consolingly. Journeys, studies, intellectual stimulation, neighbourly ties, tests and interviews and communication at all levels are all favoured in this phase of the Moon. Even taking chances in a new venture perhaps.

26 December: New Moon in Capricorn

Secret connections, the right use of contacts and friends to achieve objectives to get things done; things that have been pending for some time are satisfactorily handled now. Collaborations, perhaps even conspiracies, journeys and education – all receive a boost, an impetus.

SCORPIO

23 October–22 November

Scorpio is the sign of the Mystic, the Investigator. You Scorpios are shrewd, determined, passionate, energetic, independent, sarcastic, vindictive, intense, with much stamina, both base and noble.

I am writing this forecast on 22 January 2018. *The Times of India* of 21 January 2018 has this to say about Bill Gates, a Scorpio.

> Over time, Gates's strategy has remained mostly the same. Pick a disease that robs millions of people of their lives or livelihoods, and if a given intervention can go a long way towards reducing that suffering, chances are that Gates classifies it as 'promising'.

The two greatest moneymaking signs are Taurus and Scorpio. But Scorpio in particular is mighty keen about anything to do with health and hygiene, pollution and climate change, finance, loaning and borrowing money, and finally spirituality. Bill Gates is the perfect example.

I am writing this piece on 22 January and the film on menses has been released. Scorpio the eighth sign, speaks about sex, menses, spirituality and cleansing all that is dirty and ugly. Therefore, in a manner of speaking, the release of this picture about menses is the right release at the right time. I personally applaud it.

This time I am following it up with your digital plan for the year so that you can focus very well on it in advance and thus prepare yourself by making a blueprint of the entire year. The digital plan for the year will be short and to the point, and though it is a general reading, it should apply quite a bit to you. Health phobia and hypochondria also come under the bracket of Scorpios. Cleansing and purification are the two other specialities of Scorpio.

BLUEPRINT FOR THE ENTIRE YEAR

January: Meditation, the domestic scene, renovation and decoration, excellent rapport with people, travel and communication.

February: Home, house, family, emigration, buying/selling/renovation.

March: Top of any situation; children, hobbies and creativity.

April: Job, health, pets, projects, colleagues and relationships with subordinates and servants.

May: Love/hate, cooperation/competition, collaboration/separation, trips and ties, signing of documents and drafts.

June: Loans, funds, capital formation, buying/selling; health and vitality.

July: Start on a positive, winning streak, and journeys, ceremonies, good relationships, happy events.

August: Changes on the work and personal frontiers, need to tackle them, with tact and skill, a continuation of July.

September: Socializing, friendship, gains and wish-fulfilment.

October: Expenses, secret deals, looking after the sick and the needy, need for safeguarding own health.

November: A progressive, go-ahead month.

December: Finances, food, family, contracts and comforts.

This time we are giving you extra information about your own sign. It will be useful forever and ever. Ganesha asks for a look at the decans. Ganesha says each Sun sign occupies 30 degrees of the zodiac. The subdivisions of 10 degrees each are called the decans and reveal more details about you even though all those born between 23 October to 22 November are Scorpios.

If you were born between 23 October and 1 November (the first decan), your planetary ruler is Mars and you are a double Scorpio, with all the usual Scorpio qualities greatly pronounced in you, especially your stupendous willpower and energy but also your ruthlessness, pride and cunning.

If you were born between 2 and 11 November (the second decan), your ruling planet is Jupiter, which makes you realize your ambitions. You can sacrifice everything for a cause and carve out a name for yourself, making all your dreams come true.

If you were born between 12 and 22 November (the third decan), you are ruled by the Moon. You wish to be free and to make others free too, but you must learn to be consistent in life and not given in to melancholy and depression, nor vacillate. You are specially cut out for international collaborations, journeys and publicity.

Jupiter, the planet of prosperity as well as spirituality, will be in your second angle from 9 March 2018 to 2 December

2019. You Scorpios are very sharp and critical persons. Therefore, for you I say very simply that Jupiter in your second angle stands mainly for food, finance and family, savings and investments. Besides, the second angle also influences the following:

- Collection of piece de art, paintings, musical instruments, books, magazines, journals, manuscripts, and rare postage stamps/rare coins;
- Stepbrother or stepsister related through father;
- Risk of life or limb at the hands of spouse/lover/beloved or his mistress or her paramour or friend;
- Treaty between nations/armies or between governments of states/countries;
- Source of income and actual accrual of income to mother/stepmother;
- Receiving education in childhood up to the age of twenty years by borrowing/begging/scholarship/free-ship /returnable loan from a trust, for purposes including costs of books, writing material, clothes, food, residence, etc.;
- Enjoyment/luxury/comfort/other facilities/conveniences at the cost of others in the family or outsiders/friends;
- Gains in terms of money/prestige/publicity, etc., as compensation for defamation;
- Extracting money by dishonest/violent methods and tactic;
- Loss of money/wealth/assets by theft, fire, floods, arson,

earthquake, storm, tornado and similar other natural calamity;
- All matters relating to
  - Friend's son's wife or friend's daughter's husband,
  - Mother-in-law of son,
  - Father-in-law of daughter,
  - Renovation and decoration;
- An increase in pay, belongings, gifts, bonuses, profits, purchases and possessions.

Saturn will zoom in your third angle from 21 December 2017 to 23 March 2020 and will be important for trips, communication, computers, correspondence, contacts, signing of important deeds, papers and documents. It is a good time to appear for a test/competition/interview. The birth of a child is also possible, or you will have joy through your children or grandchildren. Most importantly, you will have the confidence so necessary for success. Saturn will help you to expand and enhance your image in the public eye, and film-makers, in particular, will be in fine fettle. Saturn, in short, favours creativity, partnerships, productivity.

According to Vedic astrology, the third angle has to do with courage and valour, physical fitness, hobbies, talent, education, good qualities, siblings, longevity of parents, tolerance, capability, quality and nature of food, selfishness, sports, fights, refuge, trading, the army, dreams, sorrows, stability of mind, neighbourhood, near relations, friends, inheritance, ornaments, cleverness, and short journeys. The third angle is also for reaching out to people and places by all mediums of publicity and transport, namely, TV, posters, pamphlets, mobile, telex, fax, Bluetooth technology, gizmos, and even by bus, car, plane, helicopter and so on. It is also about relationships, improving the vistas of your mind, the marriage of minds, so to say! It is a bonding on the physical and mental level, on the plane of ideas! Contracts and deeds will be signed. Partnerships will be essential to progress and prosperity. In short communication is the key to it all.

Your happiness quota will be 84 per cent.

SPECIAL BONUS

Ganesha says, my devotee Bejan is introducing the topic of special bonus for all of you. This special bonus is placed upon Mercury. Mercury represents journeys, ties, trips, mobility, migrations and all types of communication. Venus represents love, beauty, the fine arts, polish and finesse, diplomacy and persuasion, joy and a big hurrah to life itself. Therefore, I am combining both communication and joy for you and giving you the necessary dates.

Venus helps you in marriage and all the fine arts as well as all collaborations. Venus turns favourable between 4 February and 1 March, 27 March and 20 April, most certainly between 15 May and 8 June, 4 and 27 July, 21 August and 14 September (auspicious), 9 October and 1 November (worthwhile in every way), and finally, between 28 November and 19 December.

Mercury helps you in terms of publicity and communication. Mercury will be favourable to you from 10 February to 16 April, 7 to 20 May (very specially), 5 to 26 June (excellent for transport and journey), 19 July to 10 August (trips and ties), 29 August to 13 September (socializing and giving parties and hosting dinners), 3 October to 8 December (your wishes can be fulfilled and that is saying a great deal), and finally, 29 December to 16 January 2020.

However, I must be true to my own self. This is only a solar-scopic reading based mainly upon the position of Sun. Therefore, it may not be as accurate as a personal horoscope. At the same time, I have put in a lot of effort, imagination and inspiration as well as intuition into it. In other words, I have tried my best. The results are up to our maker. Yes, sometimes miracles do happen. Therefore, I end on a positive note, though I know my limitations very well.

KEY TO WEALTH BY NASTUR DARUWALLA

Scorpios face a lot of problem in the early stages of life. However, as they approach their twenty-eighth year, they rise so high that people watch them spellbound. The mantra for them is: *'Om hrim shrim lakshmyai namah'*. Donate blankets to the poor.

WEEKLY FORECAST BY PHASES OF THE MOON

6 January: New Moon in Capricorn

Starting the year with a bang, you Scorpios will see plenty of movement, a quickening of pace in terms of publicity ventures, journeys, changes in the domestic scene, religious rites and ceremonies, the birth of children or even brainchildren, that is, ideas and inventions. On the personal level, you might find yourself seeing the end of one relationship, followed by the birth of another, or a bonding that may be either temporary or permanent.

14 January: Moon's First Quarter in Aries

This quarter of the Moon will find you in a contemplative mood of introspection. There are bound to be sky-high expenses, overseas connections and human relationships developing. Also, care of the sick in mind and body. Once again, journeys, liaisons and rendezvous.

21 January: Full Moon in Leo

You will burst open like a monsoon downpour – and like the rain, enrich the earth and nourish it. Contracts, deeds and documents, meets, interviews and tests, solving puzzles, games requiring test of intelligence. Here you will display not only patience but also flashes of brilliant intuition and even sheer genius.

27 January: Moon's Last Quarter in Scorpio

It is transport and communication time. Once again, a time to put your ideas across the table, or even across the entire world/globe. If interested in or inclined towards TV broadcasts, seminars and conferences, drama and acting, being on a public platform, it will be a wonderful time for you.

4 February: New Moon in Aquarius

You will find yourself close in spirit and emotions not only to your spouse/mate/partner but also to people in general. You may form or break off a new relationship. Your creativity will soar; journeys with a stopover and landing plum new assignments are all likely. Thus, says Ganesha, people and places.

12 February: Moon's First Quarter in Taurus

Matters of honour, prestige, status, reputation and work/profession will be uppermost in your mind. In this context you will not only give your best, but also put yourself out to such an extent that rest and health safeguards will become imperative. Money will thankfully pose no problems – it's relaxation that will be in short supply.

19 February: Full Moon in Scorpio

The personal and financial angle is emphasized strongly in this phase of the Moon. Children, creativity and news and views all receive a great boost. People will reach out to you and you will be in great demand both socially and professionally, and you will feel yourself vibrating with life, pulsating with energy.

26 February: Moon's Last Quarter in Sagittarius

The home and family theme of the last quarter will persist. In addition, anything to do with land and development will be important. Also likely is a home away from home, a second property or holiday home. Sudden changes and development may occur.

6 March: New Moon in Pisces

The three Fs of finances, family and funds get a mighty boost. So also, your spirits, your zest for life. You will be on a social merry-go-round, with functions, meets, conferences, visits, fun at parties and picnics, all coming your way. You will have plenty of fun, and gain from it too.

14 March: Moon's First Quarter in Gemini

Group activity will continue, but work and financial affairs may see a slowing down, or some snags and hassles. It's a good time to marry – there is a strong likelihood of it anyway. I hope I'm not proved wrong on this!

21 March: Full Moon in Libra

You are again full of energy, go and enthusiasm. The artistic side of your nature will shine forth, thus the spheres of acting, parenting, investigative journalism, advertising, entertainment and love, all get a tremendous and mighty boost. You will display much finesse and subtlety and win many plaudits.

28 March: Moon's Last Quarter in Capricorn

You will be in a mood that is both sentimental and nostalgic. You will live simultaneously in the past, present and future. As the poet Shelley wrote:

> We look before and after, and pine for what is not.
> Our sweetest songs are those that tell of saddest thought.

That's you, this week.

5 April: New Moon in Aries

You will feel little vulnerable, but that will make you more human and more lovable. Lots of action this week to draw you out of the blues: house and home, the domestic scene, children, journeys, spiritual journeys and meditation.

12 April: Moon's First Quarter in Cancer

From the previous phase of the Moon till now, your angle of expenses, restlessness, spirituality, along with rendezvous and journeys, are heavily focused. Accidents and bad health may befall you, especially if driving yourself or travelling alone.

18 April: Full Moon in Libra

This phase of the Moon brings you your full share of responsibilities, duties and commitments. The good part of it all is that your efforts will be both appreciated and rewarded. Your time and energies will be claimed by the three Ps of pets, projects, policies (a wide connotation here.) To sum it all up, you will work hard and play hard.

26 April: Moon's Last Quarter in Aquarius

All through this month you have really put your shoulder to the wheel. The good results will now be apparent, and also in the coming quarter. A job hop, promotion and/or perks may materialize. You will shine forth in your chosen field, and do it with style and panache. You will gain much respect and will have better health too. Ganesha does you proud!

4 May: New Moon in Taurus

Much of what was said last week may come to pass at this time too. There may be some disruptive forces at work though, which may be reflected in marital and/or domestic squabbles either in your own family or of someone close. Whatever it is, partnerships and their dissolution will be either imminent or at the least troublesome.

12 May: Moon's First Quarter in Leo

Buying/selling/leasing will be part of the theme of improvements in house/office/shop or warehouse that is highlighted now. Your actions will invite pleasing results between October and December, though there may be some difficulties and snags now. But Ganesha exhorts you not to be dejected or give up hope.

18 May: Full Moon in Scorpio

You will be ultra-sensitive, full of your own self-respect and dignity and the desire to be loved, cherished and appreciated. You will be on a long voyage of self-discovery. On a more mundane plane, loans and funds, insurance and joint finance will be important to you.

26 May: Moon's Last Quarter in Pisces

Socializing, respectability, a good inflow of cash, and well-being will dispel the introspective mood of last week. There will naturally be more responsibilities along with more group activities and club-hopping that you may find yourself doing. A collaboration or position of trust, perhaps even marriage, may materialize.

3 June: New Moon in Gemini

You will create new sources of income, new avenues for making money in the coming years. You will build up your financial resources and clout now, and that's not a small matter. But this will be seen more clearly later during the month.

10 June: Moon's First Quarter in Virgo

Your softer and more imaginative side will come forth now. Hobbies can even become a vocation. You may develop rare foresight, telepathy and even a prophetic vision. In any case, romance, children and creative fields like art, poetry, music, literature will see you at your best. Even necromancy, raising spirits, tantra and mantra draw you.

17 June: Full Moon in Sagittarius

Money from a million different sources (I exaggerate a bit, I must confess) is what it's all about now. Loans and funds may be easy to get, lotteries and games of chance, races and the stock market may yield rich dividends. Wills, probate, legacy are also covered here. Funds and games in terms of sex and passionate encounters are likely too. Communication will have to be vital.

25 June: Moon's Last Quarter in Aries

The same direction is maintained this week. But there are many add-ons – journeys and *relationships*. Within the scope of this word come partnerships, liaisons, personal and even intimate relationships. Some bizarre situations can turn out to be to your advantage. Help from some practically unknown, and/or unexpected quarters may suddenly materialize.

2 July: New Moon in Cancer

You are off to a head start, with Ganesha's blessings. Try to slow down, pace yourself, say 'Time out'. Certainly, do not take on extra responsibilities and effort. Pets will require care. Also, you might have a theft or pilferage, or lose something valuable. Forewarned is forearmed, they say, so do take precautions. A favourable time for loans, funds and travel.

9 July: Moon's First Quarter in Libra

This quarter falls in your angle of work, job and profession, and also health. All that was said for the last quarter applies now too. Once again overexertion, strain should be avoided. Yoga and meditation, relaxation techniques and therapy, learning to de-stress yourself will be necessary.

18 July: Full Moon in Capricorn

This phase of the Moon will be a landmark in terms of growth as a person, full of knowledge, rare in sights and wisdom, that not all the wealth in the world can buy. The home and the outside world will not only meet but even clash sometimes. A trend is ushered in now that will have spin-offs and long-term benefits which will be even more apparent in 2020, and more immediately in August.

25 July: Moon's Last Quarter in Taurus

Personal issues, funds and loans are focused, also a yearning for peace and tranquillity of mind and spirit. From now till the end of August, on a material plane, renovation, decoration, buying and selling of house, home, office and vehicles will have an extra edge and require careful attention.

1 August: New Moon in Leo

You will make your mark in the world of consumer items, teaching and writing, arts and crafts, modelling, publicity and advertising, even architecture and archaeology and science. As if this was not enough, even in finance and brokerage that's how good this phase (long-term, please remember, reminds Ganesha) really is.

7 August: Moon's First Quarter in Scorpio

Health and job once again – but both on the upwardly mobile trend. People will come into your life, force you into the limelight, get work done. You will be full of charm and sociability and experience a wonderful degree of happiness along with a sense of achievement – in an area ranging from marriage to collaborations, commuting to immigration. Way to go!

15 August: Full Moon in Aquarius

This will be the time when all that's said above will probably come true, take shape. You'll find yourself automatically stepping on the gas, ready to win the race. Courage, effort, determination are all there as you volley for a winner. And win you will, says Ganesha.

30 August: New Moon in Virgo

This quarter in your own sign personalizes everything for you. You will be intense, emotional, perhaps even secretive. Mind, body, psyche are all recharged with strength and power, and of course, this well help you get things done, achieve much and strive for more.

6 September: Moon's First Quarter in Sagittarius

You are blessed with foresight, artistic excellence, joy from children, enjoyment and fun. Subtlety in dealing with problems, a certain lighting up of your being will be experienced. You will spread cheer, hope and solace, and help others.

14 September: Full Moon in Pisces

This quarter helps in finances, loans, spirituality, tantra and mantra, the occult and the unique. Buying/selling/shopping/ is also within the ambit, a total spectrum of the mind and body.

22 September: Moon's Last Quarter in Gemini

You will now hit the high spots, for sure. You're on the way to progress and achievement, with hope and love in your heart. Work and enjoyment, hard work and rest, marriage and business alliances – contradictions do meet this month and this will prevail for some time too.

28 September: New Moon in Libra

Your angle of family, finance and travel are in the spotlight. And the trio work really well for you, additionally boosted with the fine placing of Uranus and Pluto, the outer planets. You will get opportunities to make money, take a trip, acquire additional knowledge and information.

5 October: Moon's First Quarter in Capricorn

Once again, a spurt of terrific activity, in this phase of the Moon. Your imagination and inspiration power you in artistic pursuits, science and research, computers and electronics, arts and crafts and entrepreneurship. You will shine in any enterprise that requires leadership qualities.

13 October: Full Moon in Aries

This quarter makes you warm and communicative, arouses your family and filial instincts, and a strong leaning towards duty and justice. A family get-together is almost definite and you may also play a part in resolving a dispute or misunderstanding. You may take on extra work for your loved ones.

21 October: Moon's Last Quarter in Cancer

There will be expenses, travel and yet a sense of discontent alongside. More of a mood than a grievance. You may just feel that something is missing. At the same time though, this quarter of the Moon sees you creating openings for yourself. Love too is favourable and brings joy and stability.

28 October: New Moon in Scorpio

There will be an easing of tensions, of the burdens of the spirit that were troubling you, under the influence of Venus, the planet of comfort and luxury. Better health, a more positive outlook, fun and games, love and laughter, but also more responsibilities are yours. Also, confidence and conviction, meets, conferences, functions and seminars.

4 November: Moon's First Quarter in Aquarius

All that I've written just above will really become totally operational in this phase of the Moon. The ingress of Mercury in your sign now enables you to utilize the many chances that come your way. Mercury gives the power to hold, to use, says Ganesha, the Wise One.

12 November: Full Moon in Taurus

Gains and fine results, group activity and success, joy from children and luck at games of chance, easy money – these are the gifts that Mars endows you with. You have what it takes to be a winner in life. Go out and win, exhorts Ganesha.

18 November: Moon's Last Quarter in Leo

You will be resourceful, lethally accurate, and brave in the arena of life. Ambition, aggression and sheer drive will all but totally wipe out the opposition, annihilate your enemies. Parents and elders, even if no longer alive, will play an important part in your life, reaching out from the past.

26 November: New Moon in Sagittarius

Nostalgia time, time to go down memory lane, remembering long-forgotten episodes and things. On a more practical side, house, home, renovation, decoration, selling and buying will keep you busy. And what's more, you'll really enjoy it. It's part of the energized outlook initiated in the previous phase of the Moon. Once more it strikes me that the birth month of a sign is always full of tremendous activity.

4 December: Moon's First Quarter in Pisces

The world of electronics, computers, inventions – the newer, the better – will fascinate you, so much so that you become a gizmo lover! The more accurate, speedy and accurate the gadget, the more will it appeal to you. And this will continue to enthral you right into the year 2020!

14 December: Full Moon in Gemini

Riches, comforts, passion, love of fine things, beauty and art are again the gifts of Venus, for you this quarter. And richly deserved too. Even if they're not, there's a time and place for everything, and who doesn't enjoy the good things of life. Ganesha agrees.

19 December: Moon's Last Quarter in Virgo

You will be organizing functions, social get-togethers, refining management techniques. Better health will be yours too. You will work hard, and very productively, with great and sustained effort. Marriage, ties, legacies, children are all favoured. Family and money are the theme.

26 December: New Moon in Capricorn

You will be very intuitive and creatively artistic too. Some misunderstandings/tiffs may occur, but you will take these in your stride dismissing them as part of life, and succeed. Debates and angry words won't get you down.

SAGITTARIUS

23 November–22 December

Sagittarius is the sign of the Lawyer, the Sage. You Sagittarians are idealistic, honest, impatient, enthusiastic, frank and fearless, ambitious, philosophical, boisterous, argumentative and impulsive.

'Give me land, lots of land, don't fence me in'. Yes, this is the battle cry and the slogan of the Sagittarians. In other words, open space and complete freedom are the right and left eye of the Sagittarians. This year, for you Sagittarians, Jupiter, the planet of plenty and prosperity, will be in your Sun sign from 9 November 2018 to 3 December 2019. The spin-off will be a thrust in artificial intelligence, healing broken bodies, mind, and spirit (most important), managerial skills of a high order, immense strides in medicine and specially brainpower, space travel, robots, biopharmaceuticals, origin of life, God vs technology, climate change, and finally the marriage of technology and humanity. The influence of your main planet Jupiter will be over judges, lawyers, bankers, brokers, physicians, clergymen, politicians, sportspeople. The emphasis will be on higher education, the start of new undertakings, buying, lending, favours. Jupiter governs the blood, arteries, liver, thighs, right ear, sense of smell. Its metal is tin; its colours, royal purple and mixtures of red and indigo; its flower, chrysanthemum; its day, Thursday.

This time I am following it up with your digital plan for the year so that you can focus very well on it in advance and thus prepare yourself by making a blueprint of the entire year. The digital plan for the year will be short and to the point, and though it is a general reading, it should apply quite a bit to you.

BLUEPRINT FOR THE ENTIRE YEAR

January: Finances, family ties, adornment, home, buying/selling, vehicles.

February: Contacts, communication, contracts, crash courses, mental brilliance, new projects, courage and determination.

March: Home, family, treasure, parents and in-laws, work prospects even for the retired and paradoxically, impending retirement of elderly persons.

April: Journeys, ceremonies, publicity, children, hobbies creativity, therefore a lively and lucky month.

May: Jobs, pets, projects, subordinates, need for health care.

June: Love/hate, partnerships/separations, overall again a confident future.

July: Legacy, finances, passion but low vitality.

August: Distant places, research, parents and in-laws, education, children, good fortune through meeting the right people.

September: Position, prestige, power, parents, home and property, rites and rituals for the living and the dead.

October: Friendship, socializing, gains and glamour, realization of aspirations.

November: Travel, restlessness despite good fortune, expenses, need for health care.

December: Good going in terms of health, wealth and happiness.

Ganesha says this is certainly the old monthly round-up but specially for 2019 this monthly round-up becomes very necessary and precise and proper because of the planets Jupiter, Saturn and Pluto helping you in every possible way. Therefore, though it is the old monthly round-up, it still has a very special, new, different, accurate guideline specifically for 2019. Ganesha says Bejan is now eighty-seven. It is his voice of experience which says it specially for the year 2019. Let me explain to you that astrology is mainly about the right person being at the right place and at the right time and, therefore, this blueprint of the year will be of special importance and meaning to you.

Sagittarians are usually lucky in property matters. I am saying usually because my experience is limited. This year January and February, June, August and October could bring you into money. March, September and October will be important for home, house, shop, office, land, farming and so on. Please note that this is only a general forecast. I am not God.

This time we are giving you extra information about your own sign. It will be useful forever and ever. Ganesha asks for a look at the decans. Ganesha says roughly 10 degrees of the zodiac makes a decan. Since each sign has 30 degrees, there are three decans in each Sun sign. Although all those born between 23 November and 22 December are Sagittarians, the individual traits vary according to the decans.

If you were born between 23 November and 2 December (the first decan), the planetary ruler is Jupiter. You are honest and frank, with a jovial temperament and a good sense of humour. You are also both generous and compassionate. You make a lot of money easily, but spend it just as fast. Your quest for domestic serenity and for romanticism may sometimes be doomed to disappointment.

If your birthday is between 3 and 11 December (the second decan), you are ruled by Mars giving you a great zest for life and love and tremendous energy. You can be impulsive, even rash but there's nothing petty or small-minded about you. You cannot be mean, and, take on more than your share of responsibilities and duties. You have great enlightenment and recuperative powers, but have an unfortunate tendency to call a spade a bloody shovel!

If you were born between 12 and 22 December (the third decan), your ruling planet is Sun, giving you such strength of intuition that you can even be a prophet or a seer. You are made for achievement and the public eye, and are very royal and magnanimous in your interaction with other people. There is a dramatic nobility in the way you will realize your ambitions.

Jupiter, the benign and benevolent one, the giver of knowledge and wisdom, the bestower of good fortune, will be riding your sign, as per Western astrology. Let me go a little further. We Indians call Jupiter the guru, the teacher, the master, the revered one. *The Larousse Encyclopedia of Mythology* comments, 'In the name Jupiter can be found the root di, div, which corresponds to the idea of brilliance, the celestial light'. With his light guiding, you have to succeed. No other option!

Jupiter works for you in the following ways:

- Secret knowledge, journey, an initiation, travel, spiritual and uplifting experiences which just dissolve the dark corners of life and sorrow, will be yours. Yes, you will overhaul your entire life and make it grand and complete. It is like a vacuuming from within and after that, how can there be any looking back?

- *Mobile perpetuum* is the Latin for perpetual motion, and that means a grand reaching out to people and places, an ability to move, manoeuvre and manipulate, becoming precise and penetrative in letters, calls, communication, contracts, coexistence. You will interface with people powerfully and pleasurably, and that, by itself is rare and noteworthy.

- So strong is this movement and reaching out that it could result in a house or office or shop or industry move, immigration, contacts with foreign lands, neighbours, society in general. You could almost step out of your body and do some astral flying and I am not kidding! 'Move it', is the glorious and clear message of Jupiter.

Saturn the other main planet will be in Capricorn, your second angle. It will therefore emphasize food, family and finance. The other salient features will be renovation, decoration, alteration, buying, selling, leasing, shops, office, land, farm, house, in short property and assets. Architects, builders, contractors, politicians, actors, chemists, pharmacists, horticulturists, gardeners, animal trainers and even newly-weds should perform brilliantly. You will live up to the ideas of Emerson, 'To laugh often and much, to win the respect of intelligent people, and the affection of children to earn the appreciation of honest critics, to endure the betrayal of false friends, to appreciate beauty and to find the best in others, to leave the world a bit better whether by a healthy child, garden patch or a redeemed social condition, to know even one life has breathed easier because you lived – this is to have succeeded.' What is there left to say?

The two special features for Sagittarians are:

- Development of your personality, and
- Anything to do with property, house, home, parents, in-laws, architecture, buying and selling, office, shop, godown, warehouse, land.

Finally, I am combining Mercury and Venus as per Western astrology just for your entertainment, travel, parties, pleasure and children. The months for it are April, June, August, October and December. Believe me these are the months for you to enjoy yourself, help others and be happy.

The happiness quota for you is perhaps the highest of all the other signs, namely 89 per cent. In short you will be the favoured child of Lady Fortune. Pass on your good fortune to others please. That is my humble request.

This time my intuition says that the outer planets like Neptune, Uranus and Pluto will be much more important for countries than for persons; so I have not taken them into active consideration.

SPECIAL BONUS

Ganesha says my devotee Bejan is introducing the topic of special bonus for all of you. This special bonus is placed upon Mercury. Mercury represents journeys, ties, trips, mobility, migrations and all types of communication. Venus represents love, beauty, the fine arts, polish and finesse, diplomacy and persuasion, joy and a big hurrah to life itself. Therefore, I am combining both communication and joy for you and giving you the necessary dates.

For Sagittarians in particular, Mercury stands for marriage, alliance, collaborations and cooperation. Therefore, do pay special attention to the dates mentioned below:

Mercury helps you from 24 January to 9 February, 17 April to 6 May, 21 May to 4 June (very specially), 27 June to 18 July, 14 September to 2 October (many good things can happen), 9 to 28 December (simply wonderful).

Venus helps you from 7 January to 3 February, 2 to 28 March, 21 April to 14 May, 9 June to 3 July, 28 July to 20 August (very specially), 15 September to 8 October (wonderful support and help), 2 to 25 November (all pending work will be done), and 20 December to 13 January 2020 (help from unexpected quarters).

However, I must be true to my own self. This is only a solar-scopic reading based mainly upon the position of Sun. Therefore, it may not be as accurate as a personal horoscope. At the same time, I have put in a lot of effort, imagination and inspiration as well as intuition into it. In other words, I have tried my best. The results are up to our maker. Yes, sometimes miracles do happen. Therefore, I end on a positive note, though I know my limitations very well.

KEY TO WEALTH BY NASTUR DARUWALLA

Sagittarians have all the qualities of Devaguru Brihaspati. They are extroverts and friendly and attract a lot of attention. The mantra for them is: '*Om shrim hrim shrim kamale kamalalaye praseeda praseeda om shrim hrim shrim mahalakshmyai namah*'. Donate sweaters and raincoats to the needy.

WEEKLY FORECAST BY PHASES OF THE MOON

6 January: New Moon in Capricorn

You will take risks, chance your arm but hit your target! With the advent of Jupiter in your angle of money, this is a foregone conclusion. I can almost guarantee tremendous financial activity – buying/selling, shopping and investing. Hobbies and home too, and great creativity.

14 January: Moon's First Quarter in Aries

This quarter makes you contemplative, even highly introspective. An overseas angle, human relationships, care of the sick or psychologically damaged, travel, liaisons and rendezvous will all come together to make this entire month fast-paced and crucial.

21 January: Full Moon in Leo

This phase of the Moon helps you consolidate your position and make headway, get a promotion. This month will be a milestone in raking in money. New avenues will open up on the job and work scene. In family life, you will bridge gaps, heal wounds, draw close – also with siblings, kin, neighbours.

27 January: Moon's Last Quarter in Scorpio

Communication will be highlighted this quarter, but more on a personal level. Professionally too, of course, news and views, messages and calls, contracts and travel will all be both urgent and vital. A time with better things and joy awaits you.

4 February: New Moon in Aquarius

Your job, profession, business, career are accentuated this quarter. It's a good time to be generous, give and receive favours; and you will be sought out for advice too. Health could pose problems. Balance and maturity are necessary at this juncture in professional relationships, especially with your boss. You will balance melancholy with joy, be introspective and somewhat withdrawn, with a deep stillness within you.

12 February: Moon's First Quarter in Taurus

The personal and financial angle will be strongly emphasized. Children, creativity, communication (in terms of news and views) – the three Cs get a strong boost. So do your libido and feelings. People will reach out to you for love, work, assistance. You will come alive, vibrantly.

19 February: Full Moon in Scorpio

This phase of the Moon will galvanize you into action, especially in the spheres of finance and communication. You'll have to take quick decisions and move fast. Your actions will have an impact in March, so will the trends of January that is buying/selling/trading and so on.

26 February: Moon's Last Quarter in Sagittarius

A little rest, relaxation and some health care will prove to be a must. Ambitious and hard-working as you are, you tend to overdo things. Elderly Sagittarians please note, you may feel prone to depression and here's where mind, body, healing, yoga and meditation will hold a tremendous appeal.

6 March: New Moon in Pisces

Job opportunities will multiply, or maybe there is an additional assignment at work. Anything to do with property and house, land, building/showroom/warehouse/shop will be in strong focus. Older people will have health problems, perhaps even contemplate retirement. Your parents/in-laws/boss could also cause anxiety about their health.

14 March: Moon's First Quarter in Gemini

You will extend hospitality, warmth and generosity, not just help. There may be the break-up of a tie, but you will soon bounce back. So don't brood over it, advises Ganesha. In any case the fine placing of Uranus and Pluto makes it a good time for material success and emotional bonding. People will respond to you.

21 March: Full Moon in Libra

You will open up power full-throttle now. And there's enough to send you into a mad spin this quarter, and month! Messages and links, news and views, trips and relationships (with in-laws, relatives, visitors too) all make you feel that you'll never be able to cope. But cope you will, in fact even laugh and enjoy it. Also, a phase in your spiritual and emotional life will conclude now.

28 March: Moon's Last Quarter in Capricorn

A mood of nostalgia and sentimentality now, even looking

back to the past. You may feel vulnerable, open to hurt but it certainly makes you more lovable. House and home require attention, so also children, the domestic scene and friends. Journeys are foretold too, by Ganesha.

5 April: New Moon in Aries

You are very human, easy to love and you will be loved too. That's wonderful. People will reach out to you as you reach out to them, and sudden and unexpected help from friends, and well-wishers will warm and gladden your heart. Buying/selling/shopping, especially for the house/home/office are what you're chiefly involved with in this quarter.

12 April: Moon's First Quarter in Cancer

Creativity and pleasure for you. You will be in a strangely – and for Sagittarians – uncharacteristic, eccentric mood, even contradictory. You will be both radical and conservative in your views and opinions. It will also be a landmark in your growth as a person, and your understanding of life and living. The influence of this will spread over more than the next two years.

18 April: Full Moon in Libra

You will be highly productive and creative in this phase of the Moon. Power and responsibility, promotion and perks – you really do well. Your bank balance and credit will need to be examined and your home too will need attention and care.

26 April: Moon's Last Quarter in Aquarius

The good results of all your efforts and hard work will now be seen. You will make it in style in your chosen field. A job hop, and once again promotion and perks are emphasized. You will gain in respect, have better health, and put your best foot forward.

4 May: New Moon in Taurus

You will attempt the impossible, take great risks, even throw caution to the winds. Creativity and romance are the watchwords. It's the right time to start an enterprise, but especially for poets, writers, musicians, researchers and teachers, chemists and scientists, organizers and entrepreneurs, actors and doctors. The list, though long, indicates all those who earn by the intellect rather than the sweat of their brow.

12 May: Moon's First Quarter in Leo

Once again, restlessness, introspection, innovation and bright ideas are Ganesha's gifts. Plenty of opportunities to make money too. Distant places attract you, original thoughts come to you, spiritual pursuits and meditation draw you. Your brain cells are not just activated but working overtime! Ganesha be praised.

18 May: Full Moon in Scorpio

One of the grandest of phases. Ganesha grants a wish-fulfilment, getting what you want and where you want – right through till mid-June. Sports, vacation, travel, comfort, the joys of friendship are yours to enjoy. People with artistic talent will be especially productive, successful, earning both recognition and satisfaction.

26 May: Moon's Last Quarter in Pisces

Communication and travel are very favourable for you. All types of media and methods of information and entertainment will be used to definite advantages by you. Tourism and travel-related ventures will be profitable too.

3 June: New Moon in Gemini

There's much for you in this quarter of the Moon. Socializing, a sense of well-being, respectability, a great surge in finances, also clubbing and group activities. All this will naturally bring more responsibilities. Also favoured are collaborations, a permanent alliance, responsible position and marriage. Perhaps marriage sums up all the other three.

10 June: Moon's First Quarter in Virgo

Once again, introspection and restlessness claim you. On the other side, the pendulum swings towards foreign connections, travel to distant places, liaisons, research and investigation. You will cope with this paradox, even though June promises to be a slightly difficult month.

17 June: Full Moon in Sagittarius

This phase of the Moon brings more money, from a variety of sources. In fact, no matter how materialistic it may sound, let's be practical and so it's my sincere advice to all Sagittarians to concentrate solely on making money—oodles of it, in fact. Sources other than honest toil will yield money: lotteries, legacies, races, games of chance. You are vibrantly, gloriously alive and passion and sex are part of it.

25 June: Moon's Last Quarter in Aries

All of the above will get a further nudge in the right direction this quarter, which favours partnerships and journeys, rendezvous and personal relationships. Yet again, help from an unlikely/unexpected source materializes for you. A truly bizarre but beautiful week. Great fun, chortles Ganesha.

2 July: New Moon in Cancer

This phase of the Moon finds you efficient and organized, clearing away all the dead wood. Not just ambitious, but raring to go. Impatience and intolerance of other's failings *must be avoided*. Learn to make allowances. Exceptionally heavy responsibilities may accompany a promotion. On the personal front, show people you care, at least your loved ones, because care you do and deeply.

9 July: Moon's First Quarter in Libra

Your job and health angles are in focus of this week. Learn to go slow. Paradoxically, when most people must be pushed to work hard, Sagittarians as a rule have to be reined in. Pets will require care, and loss, theft, misplacing of valuables has to be guarded against. This month will, however, be favourable for loans and funds, and travel is almost a certainty.

18 July: Full Moon in Capricorn

This phase of the Moon gives you a kick-start, a fresh impetus, making you extremely sensitive to life itself. You will be keener and sharper at work; in fact, I can extend this to say you will be dexterous in all skills. Not only this, great personal progress too is guaranteed, if you can stop daydreaming, building castles in the air and giving in to wild impulses.

25 July: Moon's Last Quarter in Taurus

You may be in a gay, sportive, playful mood in this quarter–rare for Sagittarians. Even if you don't feel you're sprouting wings, you will feel on top of the world. Sports and hobbies and children naturally reflect this ebullience. So also may speculation, financial dealings, loans, love affairs, music and the arts. A wide spectrum indeed!

1 August: New Moon in Leo

This quarter looks at the August theme, which will be one of immense productivity. Your delightful intoxication with life persists too. You will be like a king or queen. Creative pursuits reflect this joyousness too.

7 August: Moon's First Quarter in Scorpio

An ideal time not only to go out but forward as well. You will please not only yourself but also others. Once again, artistic pursuits and creativity are favoured and also any form of business, trade or entrepreneurship dealing with the arts. Communications and contacts will flourish too.

15 August: Full Moon in Aquarius

This will be decision time for you, when you go all out, pull out all the stops in your pursuit of your goals. Increased finances and emotional satisfaction will both be there, to keep you happy, fulfilled and easy in mind and spirit. It is a good time for loans and funds and wonderful to socialize, have friends over.

30 August: New Moon in Virgo

This quarter puts everything on a personal level for you,

makes you intense, secretive, emotional. You will have the power not only to get things done, achieve objectives, but build up inner reserves of strength and spirituality.

6 September: Moon's First Quarter in Sagittarius

The full Moon phase has much for you – intuition and foresight, joy from children, a sense of fun, artistic excellence. Love enriches you. You will experience a definite lightness of being and will spread cheer and hope all around you, untangle knotted problems with insight.

14 September: Full Moon in Pisces

Gains through buying/selling stocks and trades are foretold. Extra attention to renovation/decoration or sale/purchase, of house, office premises, warehouse are also likely. Comforts and luxuries for the home or office, the latest in labour-saving devices and white goods will be required. You may even visit auctions and flea markets to pick up interesting bargains and artefacts.

22 September: Moon's Last Quarter in Gemini

This quarter of the Moon will be extremely productive. You may be well on your way to acquiring both wealth and fame. Calls and correspondence, trips and commuting are all likely. All matters related to finance, marriage, travel, advertising and publicity, news and views, will contribute to your sense of well-being and financial betterment.

28 September: New Moon in Libra

A phase full of red-letter days – ceremonies, collaborations and mergers come your way. In addition, you will be active and inspired, almost prophetic in your vision, outlook on life and attitudes to living (there is a difference), so too with travel. A time to complete pending work as well.

5 October: Moon's First Quarter in Capricorn

This quarter focuses strongly and favourably on your angle of work, and the rewards and awards connected with it. Perks, a fatter pay packet or money from unexpected sources, a bonus or gift may come your way. The emphasis, however, remains firmly on *rewards for efforts*.

13 October: Full Moon in Aries

You will 'win friends and influence people' effortlessly. You certainly don't need to read Dale Carnegie. Care of children, house, home, family gatherings, entertainment and amusement all give you an opportunity to shine, to burnish your personal image. Consumer goods and machinery (their correct use and packaging, to be more precise) are favoured too.

21 October: Moon's Last Quarter in Cancer

A strange disenchantment grips you, a mood in which you might quote Shakespeare, like me, and say, 'Now is the winter of our discontent.' Prayer and meditation could heal you. Work, thankfully, does not suffer; in fact, the influence of Mars could help create a few new openings. Love will be the rock on which you build. This phase also brings expenses, travel and some behind-the-scene activities.

28 October: New Moon in Scorpio

You swing back with a vengeance. The astrological reason

is there. You get a second chance to make good, and are likely to come out right on top. Personal affairs could prove a little troublesome, especially if you are sentimental or oversensitive. You have to be direct and pragmatic in your approach.

4 November: Moon's First Quarter in Aquarius

The social scene may claim you – having a good time, making money, joining clubs, replenishing your wardrobe. It helps you cope with the swings between extroversion and introversion that you are prey to. You will display natural magnetism and look inwards too.

12 November: Full Moon in Taurus

From the last quarter onwards, you are likely to be hurled into terrific activity, which carries over into this quarter too. Socializing, teamwork, strong and intense friendships, and also fresh means of making money all conspire to keep you really busy. But the good thing is that the results will show soon.

18 November: Moon's Last Quarter in Leo

You will be occupied with a number of things and events in this action-packed quarter. You will hit the big times, in the three Cs of contacts, commuting, communication, even conversation (interaction with people). Journey, ceremony and publicity are favoured. Also, the likelihood of a marriage or engagement. Ganesha helps you to cope.

26 November: New Moon in Sagittarius

Emphasis strongly on financial activities now. Loans, funds, additional sources of income, buying, selling, shopping, trading keep you busy. Those dealing with the spoken or written words will do well and gain more.

4 December: Moon's First Quarter in Pisces

An irresistible fascination with the latest in gadgetry will possess you. The more fancy and efficient the gizmo, the better, as far as you are connected. Your love of precision, of making things work, makes you appreciate all this, also the electronic media, the latest in computers and so on.

14 December: Full Moon in Gemini

This quarter focuses on personal matters. A lot of activity and coming and going are predicted. To spell it out, there could be marriage, the affairs and concerns of children, handling of conferences and meets, attending or arranging functions! A hectic time warns Ganesha.

19 December: Moon's Last Quarter in Virgo

You will be working at a place that defies description. Not just work, playing hard too, and at the same place. There will be strong overseas connections – distant lands and foreign places. You will feel you are now getting what you deserve, what you have struggled to achieve.

26 December: New Moon in Capricorn

Creativity at white heat, artistic excellence and remarkable intuition for you this quarter. Also, some quarrels, tiffs, debates or words spoken in anger. It's a good time to curb your natural impatience and not give in to hot-tempered retaliation. Upsets are part of life, and have to be coped with.

CAPRICORN

23 December–22 January

Capricorn is the sign of the Priest, the Administrator. You Capricornians are hard-headed and practical, ambitious, suspicious, resentful, diplomatic, reserved, selfish, orthodox, determined and unscrupulous. Ganesha says according to Western mundane astrology, India's main planet is Saturn. Saturn means ambition, mighty spirituality, old religion, Mother Nature, duty, responsibility. But Saturn also means very fixed old ideas, great difficulties in accepting change, fixed opinions and values, sometimes pettiness. Both Dhirubhai Ambani and Ratan Tata are Capricornians born on the same month and date. That says it all.

This time I am following it up with your digital plan for the year so that you can focus very well on it in advance and thus prepare yourself by making a blueprint of the entire year. The digital plan for the year will be short and to the point and though it is a general reading, it should apply quite a bit to you.

BLUEPRINT FOR THE ENTIRE YEAR

January: Success, happiness, fun and games, family, victory.

February: Finances, family, the luck of the draw, buying, selling, investing.

March: Fanning out to people, places, contracts and contacts, communication channels.

April: Home, house, parents, in-laws; renovation/decoration of office/shop/home.

May: Enjoyment in journeys, creativity; joy from children.

June: Work, health, loans, pets and projects.

July: Off and away to a magnificent start, despite pressures and pulls.

August: Windfalls, joint finances, legacy, passion.

September: Right contacts, success, travel, publicity.

October: Tremendous drive, like Virat Kohli, perfectly honed ambition.

November: Good news, pressures, delays, turning out well in the end, socializing.

December: Expenses, losses and psychic powers, glimpses of Ganesha/God/Allah/the Supreme Power, pilgrimages, rites and duties.

Ganesha says this is certainly the old monthly round-up, but specially for 2019 this monthly round-up becomes very necessary and precise and proper because of the planets Jupiter, Saturn and Pluto helping you in every possible way. Therefore, though it is the old monthly round-up it still has a very special, new, different, accurate guideline, especially for 2019. Ganesha says Bejan is now eighty-seven. It is his voice of experience which says it specially for the year 2019.

CAPRICORN

Let me explain to you that astrology is mainly about the right person being at the right place and at the right time and therefore this blueprint for the year will be of special importance and meaning to you.

You Capricornians are certainly very efficient, and have great managerial skills, and while you could be ruthless, it is also a given fact that you specialize in service to humanity. Swami Vivekananda was a Capricornian. Therefore, there might be contradictions in you, but it is certain that you can be of great help to one and all. Therefore, please do not take the digital plan lightly. I have worked hard at it.

This position of Saturn in Capricorn ensures success in the materialistic, financial or fame sense of the world, and it will not long be satisfied with unrecognized worth. It is the master key.

Ganesha asks for a look at the decans. Ganesha says each sign of the zodiac is 30 degrees and when further subdivided into 10 degrees each they form the three decans within that sign. Thus, although all those born between 23 December and 22 January are Capricornians, their individual traits vary according to the decans.

If you were born between 23 December and 1 January (the first decan), your ruling planet is Saturn, and it is very powerful and at its maximum operative force. This is an ideal decan for engineers, industrialists, contractors, empire builders and politicians. Anything connected with land and houses is to your ultimate financial advantage. A resigned, stoical attitude is also seen.

If you were born between 2 and 11 January (the second decan), your planetary ruler is Venus and you combine utility and beauty successfully which yield both pleasure and profit. Frequently, those born in this decan have emotional or marriage problems, but it is probably the most 'balanced' of the three decans.

If you were born between 12 and 22 January (the third decan), Mercury is your ruling planet, making you reach out to people. Use your contacts and influence well. News, media, communications not only just interest you, they shape your life and your thinking as well as your choice of profession and success in it, in a safe and concrete way.

This is what Jupiter will do for you. Ganesha claims you will stand 10 feet tall, as you have 'grown' in every possible way imaginable in 2019. The forward thrust you received in your activities and attitudes now prepares you for a lot of movement, and yes, it is movement which becomes the catalyst and motivator of your life.

What does this movement actually mean?

- It means a journey, a house or office move. That could, in turn, have far-reaching consequences.

- Thanks to this movement, hidden things or memories come to life. Even as you advance – and advance you will – the past becomes vivid and springs to life.

- Your spiritual/religious tendencies and inclinations will surface, and if your personal horoscope is strong, there is a distinct possibility of psychic abilities and great intuition, as well as prophetic dreams, being fully manifest.

- I am prepared to bet on one aspect – that you learn a lot about yourself, through introspection, circumstances, relationships, psychology, psychiatry, travel, reading, interacting with nature, society and also your own self. It is a period of self-knowledge and self-discovery.

- Should you desire to immigrate, or have strong foreign or long-distance connections, 2019 is a pretty good year for it, despite the poor formation of Jupiter and Saturn, according to Western astrology.

- The softer, more humane side of your personality will come to the fore. You could be a healer, a teacher, or find one. The astrological reason is that Jupiter stays in your hidden or secret angle till then. And after 2 January you will have learnt all your karmic lessons and, thus, be ready for the goodies which life will very definitely lay at your feet. We will talk of that at leisure and at length when we come to 2020, the next year.

In the tapestry of 2019, the design and colour which will be very sharp and clear takes the form of contacts, communication, correspondence, contracts, commuting, the five Cs, as I call them. I know and understand, that, to all of us, these five Cs are important. But in 2019 and to the Capricornians, it will make the difference between a beautiful and useful life and mere survival.

Ganesha adds, the creative and constructive inclinations of Capricorn will find happy outlets in teaching, writing, administration, spirituality, arts, research. Feelings will be deeply stirred, rather unusual for you folks, but the experience will be worth it. According to Western astrology Saturn will be in your sign Capricorn from 21 December 2017 to 23 March 2020. This is indeed a long run for Saturn. The actual results will be:

- Older people will retire but possibly find new work.

- There will be enormous social changes in the form of new ideas, innovations, inventions and a mighty struggle with old conventions, rigid customs and traditions and manners and actual way of thinking. In other words, a great change in the very behaviour pattern and thought processes of the people of the world including India. It is a time of transition and transformation but the final result, according to your Ganesha devotee, should be in favour of science, technology and new as well as different modes of thinking. Thinking leads to behaviour. Behaviour leads to action. Action leads to change. Therefore, like it or not, we are in for a big revolutionary change in our thinking. Finally, it will help us in our own evolution. We often forget that everything in life has a price. Nothing is free. Evolution could be painful but necessary and sometimes inevitable. At the same time, I openly agree that I could be hopelessly wrong. I am guided by my Lord Ganesha, Mother Ganga and Mother Nature. But I openly admit that being a human being I have my limitations.

One of the salient features will be a very active, positive and well-planned move for space travel so much so that we could be preparing for migration within three years. But for actual migration perhaps the year from 18 December 2020 to 8 March 2023 will be of utmost importance. I am informing you about it because I do not believe I will live long enough for you to point out about this to you. This is only an insight born out of Ganesha's grace, and my intuition and hard, bitter and sweet experience of over seventy years. But let me clarify that in life nothing is final. That is why life is always exciting and not completely predictable. That is the way I like it.

Let me summarize it for you. Ambition + efficiency + drastic changes in climate + human behaviour + structure + technology = a new and wonderful spirituality as well as humanity.

Your months of pleasure and profit will be March, May, September and November. Go for it. I know that you Capricornians are resolute and determined. Therefore, your chances of coming out on top are excellent. This is certainly my last word.

Your happiness quota will be 83 per cent.

SPECIAL BONUS

Ganesha says my devotee Bejan is introducing the topic of special bonus for all of you. This special bonus is placed upon Mercury. Mercury represents journeys, ties, trips, mobility, migrations and all types of communication. Venus represents love, beauty, the fine arts, polish and finesse, diplomacy and persuasion, joy and a big hurrah to life itself. Therefore, I am combining both communication and joy for you and giving you the necessary dates.

For communication and coordination Mercury helps you from 5 to 23 January, 10 February to 16 April, 7 to 20 May (very definitely), 5 to 26 June, 19 July to 10 August, 29 August to 12 September (a grand period), 3 October to 8 December (you finish the year on a winning note)

Children, hobbies and creativity all come under the planet Venus very specially for you. Venus helps you from 1 to 6 January, 4 February to 1 March, 27 March to 20 April, 15 May to 8 June (very specially), 4 to 27 July (it will blow hot and cold), 21 August to 14 September (possibly wish-fulfilment), 9 October to 1 November (loved ones hug you and love you), 26 November to 19 December (you can be all that you want to be). That is saying a lot.

However, I must be true to my own self. This is only a solar-scopic reading based mainly upon the position of Sun. Therefore, it may not be as accurate as a personal horoscope. At the same time, I have put in a lot of effort, imagination and inspiration as well as intuition into it. In other words, I have tried my best. The results are up to our maker. Yes, sometimes miracles do happen. Therefore, I end on a positive note though I know my limitations very well.

KEY TO WEALTH BY NASTUR DARUWALLA

Capricornians have great intelligence. They are known for their patience and hard work. They hardly get affected by things like hurt or criticism. The mantra for them is: '*Om shrim hrim klim aim saum om ka a ee la hrim ha sa ka ha ka hrim sakal hrim saum aim klim hrim shrim om.*" Donate medicines to the needy and help them undergo regular check-ups if possible.

WEEKLY FORECAST BY PHASES OF THE MOON

6 January: New Moon in Capricorn

You start the year with a bang, pushed into the limelight of publicity. You will have more energy than you know what to do with. You will score heavily, and work with a will. Good things, therefore, are bound to happen to you.

14 January: Moon's First Quarter in Aries

You will be imaginative, bold and daring in your approach to life. At the same time, organizing skills will be displayed.

Love too warms the cockles of your heart this frosty month. Matters related to house and home, partnerships, travel with a stopover – all find you at your best.

21 January: Full Moon in Leo

This phase of the Moon will help you consolidate your position, make headway, gain in promotions and perks. Messages, news and views, contacts and communication will be urgent and require attention and care. Capricorns are good at this.

27 January: Moon's Last Quarter in Scorpio

I call it money and honey. Please add on travel, contacts, confidence and a good deal of fun, and you get the picture. Before I forget – partnerships too from the purely personal to the purely platonic. A wide range!

4 February: New Moon in Aquarius

This month has been a milestone in making money and for work, yet there has been an even tenor, a placidity that was missing earlier. It is this that is strengthened this quarter – reflected in all you say, do, pursue. Your self-esteem will rise, gains will be consolidated, particularly financially.

12 February: Moon's First Quarter in Taurus

Friends, favours granted, love, the good things of life for you. There is a rare coming together of pleasure and profit, and plenty of entertaining and socializing. Those related to the arts, but most particularly music, show exceptional flair and do admirably.

19 February: Full Moon in Scorpio

Finances and communication are triggered off tremendously during this phase of the Moon. Most of your activities of now will have spin-offs in March. It's time to think on your feet, take quick decisions and act on them equally fast.

26 February: Moon's Last Quarter in Sagittarius

A phase when you can collect on favours done, accomplish your objectives, and share your good fortune with others. Promotions may materialize, but on the whole, a week to enjoy yourself, socialize, have a good time.

6 March: New Moon in Pisces

A truly beneficial phase – with the fine placing of Sun and Saturn you will come out on top in tests, trials and competitions. It has also got the best scope to earn favours and receive good news. A little health care will be a must, also relaxation to recharge your batteries.

14 March: Moon's First Quarter in Gemini

An ideal phase in terms of contacts and communication – for both pleasure and gain at work. Also, for publicity, and socializing. It's fine time for you steady and sincere Capricornians to socialize and lighten up – both yourselves and the environment you live in.

21 March: Full Moon in Libra

A crucial time, in terms of being favourable for generally making headway. More specifically for important deeds and documents, for paperwork and proposals and tenders. Ganesha warns you to read the fine print carefully. There can be snags that would bother you later. The astrological reason is that if you overdo things, you may not get the results anticipated. That is the effect of Saturn being placed with the Sun, as I mentioned earlier.

28 March: Moon's Last Quarter in Capricorn

Trips and ties sum up this quarter. Once again, a word of warning. Accidents are not ruled out. Expenses will be there, and some kind of waste of money. In that context, air travel might not give all the benefits you'd expect.

5 April: New Moon in Aries

Yoga and meditation may help you cope well with the strains on you over the last few weeks. A stunning understanding of people and deep inner maturity will come to you. This, ironically enough, also comes from Saturn. You will be a great counsellor and arbitrator.

12 April: Moon's First Quarter in Cancer

Partnerships at all levels, and both personal and professional, are focused this quarter. You will be in the thick of corporate battles, struggles to get your point across, jostling for power and position. Don't go too far on the aggression – it might rebound on you. It's always wise to give a little, take a little, advises Ganesha.

18 April: Full Moon in Libra

As hinted earlier, your personal and professional images will be in strong focus. It is a good time to watch your step, take precautions and safeguards. Your motives can be misunderstood, and your intentions misconstrued. Play it cool is good astrological advice.

26 April: Moon's Last Quarter in Aquarius

You will be enormously creative and productive. The previous phase of the Moon ushered in a change which will be more apparent now. If you move with restraint and caution, promotion and perks will be yours.

4 May: New Moon in Taurus

The theme of pleasure and profit continues. All the good things of life come your way. For illustration, the latest gadgets, good food, music and laughter, and beautiful things for the home. And it's all made possible by your own hard work and enterprise.

12 May: Moon's First Quarter in Leo

New energy, new life, which will actually be more visible in the next quarter, start manifesting themselves now. It's a time of great hope, new perspectives. Some things which have bothered and distressed you will start easing up now. There's light at the end of the tunnel for you.

18 May: Full Moon in Scorpio

All the trends hinted at last week become crystal clear in this quarter of the Moon. You get what you want. Ganesha has granted you wish-fulfilment. Travel and tourism yield pleasure and profit, so do sports, vacations, artistic endeavours. You will enjoy all the comforts of home and the joy of friendship.

26 May: Moon's Last Quarter in Pisces

This quarter offers excellent opportunities for advancement for those already in high position, especially in the fields of arts, teaching, management. They'll feel they've never had it so good. It's a hard fact of life that those who earn by their intellect seldom get their just material rewards. But at this time, you do.

3 June: New Moon in Gemini

You will be riding the crest of a wave, surf-boarding your way to success and recognition as this lucky path actually lasts till mid-June. You will find grand things happening to you, especially in the spheres of general happiness and well-being, romance and speculation.

10 June: Moon's First Quarter in Virgo

Once again introspection and restlessness. You are never one of those who, flushed with success, take the good times for granted. Religion, research and study, foreign connections, liaisons and love affairs, and also travel are all favoured this week. So you will actually be performing on two levels of consciousness.

17 June: Full Moon in Sagittarius

The foreign angle and overseas connections will be further activated now. A house move, living abroad, or travelling there, export–import are all favoured. There will be plenty of movement and your intuition and sixth sense will guide you well.

25 June: Moon's Last Quarter in Aries

Work will suddenly override practically everything else. You were never one to shirk work, so you may be putting yourself flat out. Now's the time to learn relaxation techniques, mind and body healing, exercise and general health care to avoid trouble later.

2 July: New Moon in Cancer

The general drift of the last quarter forecast holds good this week as well. Harder and longer hours of work, a likely job switch or a job for the jobless. The pressure will be immense, but so will the potential opportunities. But, says Ganesha, you must balance between the two.

9 July: Moon's First Quarter in Libra

A new focus emerges in July, which may continue right till September. The emphasis will be on links and ties, the basic emotions of love, and its counterpart, hate. You will run through the whole gamut of human relationships, starting now. Some opposition or upsets may occur. Body and mind therapy, spiritualism, even religion; or at least New Age thinking, if that's your bent, will help.

18 July: Full Moon in Capricorn

All that I wrote in the paragraph just above will be thrown into strong focus this quarter. And you will be thrown willy-nilly into all the above, plus monetary transactions, loans and finances, buying and selling and trade, as you realize that is where better prospects for the future lie.

25 July: Moon's Last Quarter in Taurus

There will be an addition to your preoccupations with alterations in house/home/office, some upsets or snags or differences of opinion too. Surprisingly, you will get a lot done in July even if you have to struggle and tussle to achieve it.

1 August: New Moon in Leo

A real medley of home affairs and work, attraction and being repelled or put off, travel and a desire to stay put. But contradictions do meet, as John Donne said, 'All you have to do is give a little in terms of leeway, consideration, and compromise.' Don't be hard on yourself or on others – that is the astrological advice from Ganesha.

7 August: Moon's First Quarter in Scorpio

You will be openly expressive and vocal this quarter, and will certainly pull no punches in debates and arguments. The good thing is the softening impact of love. Children, romance and good money add to it. You learn to be grateful for all you have, to count your many blessings.

15 August: Full Moon in Aquarius

This quarter is advantageous for you in the world of money – that's supposed to be the second most important thing in the world, though some misguided souls put it first! Loans, funds, joint finance, interest and rentals are all favoured. Also, trade, buying and selling, brokerage and new assignments. You'll need to pull out all the stops to cope!

30 August: New Moon in Virgo

The focus is on partnerships, whether for business/profession or an official engagement or wedding. I am compelled or use another of my favourite phrases to sum it up – money and honey. They can both be there for you, thanks to Ganesha.

6 September: Moon's First Quarter in Sagittarius

The full Moon sends you off on journeys, and voyages of communication, a major move. You will acquire a different perspective on life and living. A very bizarre combination of love and libel suits may also have to be dealt with. How, I don't know!

14 September: Full Moon in Pisces

The impact of the previous phase will extend over this quarter too. Your work and efforts will show results later in the quarter, and rewards and promotions may come to you. You may feel they were long overdue.

22 September: Moon's Last Quarter in Gemini

Once again, action time, being productive time, getting places time. Finance, marriage, advertising and visual publicity, news and views, pave the way for riches and fame for you. The other emphasis is strongly on contacts and communication, even if it means a certain amount of travel.

28 September: New Moon in Libra

An extremely unusual quarter. You will make all the right moves, inspired by an inner voice, somewhat like Joan of Arc's divine voices. Strangers, foreigners and visitors will be helpful. Travel will be beneficial and love affairs will gladden and excite you.

5 October: Moon's First Quarter in Capricorn

Once more, house, home and family (particularly parents and older relatives) will all demand care and attention. And once more, the home and family theme will carry over into October. Buying or selling or renovating property is likely. This may even be linked to prospects of retirement for older Capricornians, though here some form of re-employment may be a possibility.

13 October: Full Moon in Aries

Once again, home and family matters along with contacts and alliances. Relationships will prove to be the salient features of this phase, especially marriage and divorce. Travel or a journey with a stopover is likely.

21 October: Moon's Last Quarter in Cancer

You've now to push ahead with all you've got, make all possible efforts to get on in life. Good things and benefits must be fought for, and if you don't, others are waiting to get there instead. You may need to be devious, clever, even crafty to achieve your objectives, especially in money matters. And remember, this is a monthly trend, so some rethinking and realignment of priorities will be necessary.

28 October: New Moon in Scorpio

How many of us are given second chances in life? You're some of the lucky few to get a fresh opportunity to make good. And Ganesha lays odds on your winning. Personal affairs may give you a rough time, but learning to impose curbs on yourself vis-à-vis expectations and sentiments will help you tide it over.

4 November: Moon's First Quarter in Aquarius

This quarter of the Moon is a truly decisive one and sees you achieving a real turnaround in your affairs. Romance and finance (I have this weakness for pairing things, or clubbing them together) are specially favoured. The satisfaction and effects will be savoured and appreciated later in this month.

12 November: Full Moon in Taurus

Myriad possibilities open up for you in this quarter. I might even say the entire scope of your activities responds to the astrological stimulus. It is for you to assign priorities, exercise management and PR skills. Job and travel, or perhaps a combination of both, are favoured, along with financial deals and a touch of romance.

18 November: Moon's Last Quarter in Leo

The highlight remains finance. You might well find money pouring in shovelful by shovelful, if you've made the right moves so far. If not, there's still hope, there's still time as this is an immensely lucky phase and you might just make it in the race to the top. You will be generous, loving and caring with all this.

26 November: New Moon in Sagittarius

This phase of the Moon starts with a fine trend. You may well see its culmination now. You'll give and receive in large measures. I'm not only referring to money but certain intangibles too. Bygones will be bygones and love and harmony will shine forth. Hallelujah!

4 December: Moon's First Quarter in Pisces

Finances are immensely favoured once again. You will realize your true potential and promise in many spheres, not least of which will be personal relationships, friendships and human relationships. All this is partly due to the planets and partly the rethinking of values on your own part.

14 December: Full Moon in Gemini

A time that is tailor-made for interfacing, interacting, human relationships. Proposals of marriage or a love affair may materialize. It's certainly a wonderful way to interface after all. Jokes apart, working as a team, pulling together, are strongly emphasized by Ganesha.

19 December: Moon's Last Quarter in Virgo

This quarter favours journeys, and physical or mental moving ahead. In this will come crash courses, intensive study and research, tests and interviews mental stimulation, intellectual growth of all kinds. Ties with neighbours and the theme of communications too.

26 December: New Moon in Capricorn

The previous phase also favoured secret activities. It will now extend in this quarter to pulling strings, using the right contacts and maybe even secret connections to achieve your objectives. A conspiracy or secret plan may even be hatched. Please note the repeated use of the word 'secret' in this forecast, and interpret and act accordingly.

AQUARIUS

23 January–22 February

Aquarius is the sign of the Inventor, the Seeker of Truth. You Aquarians are honest, popular, amiable, truth-seeking, hesitant, inefficient, humane yet suspicious, rebellious, intuitive, broad-minded and unbiased. Ganesha says the sign Aquarius as per Western astrology stands for the actual Age of Aquarius. Aquarian means Kumbha rashi by Vedic astrology. Therefore, the Age of Aquarius is speaking of the old Kumbha mela. But there is a difference. In the Kumbha mela people come together and they do so by using the five Cs, namely, connection, communication, collectivity, contacts and creativity.

This time I am following it up with your digital plan for the year so that you can focus very well on it in advance and thus prepare yourself by making a blueprint of the entire year. The digital plan for the year will be short and to the point, and though it is a general reading, it should apply quite a bit to you.

BLUEPRINT FOR THE ENTIRE YEAR

January: Despite expenses and interferences, property matters, and family conditions do give some satisfaction; buying/selling, journeying are emphasized.

February: Go all out for the kill, roaring like a lion, and emerge victorious in whatever you do.

March: Finances and funds will be augmented.

April: Contacts, contracts, socializing, friendship, good news, and you will make your presence felt.

May: Home, house, family, parents, property, renovation/decoration, buying/selling, leasing, shopping are indicated.

June: Plenty of fun and frolic; children add joy to your life, hobbies give deep satisfaction, sports and creativity fulfil you – a great ending to a busy beginning, concludes Ganesha.

July: Work and projects could tell on your health unless you learn to relax.

August: Love/hate, attachments/separations, journeys, a home away from home, marriage/divorce – all these are emphasized.

September: Funds for work/home, hurts, buying/selling/investing, capital formation are all indicated.

October: Journeys, publicity, exaltation and execution, collaboration, a grand reaching out to people and places, all these will keep you involved.

November: Work is worship could well be your motto; for good measure, add duty and beauty.

December: Love life, laughter and the law of chances operate in your favour. So if you feel like it, take a few chances.

Ganesha says this is certainly the old monthly round-up, but specially for 2019 this monthly round-up becomes very necessary and precise and proper because of the planets

Jupiter, Saturn and Pluto helping you in every possible way. Therefore, though it is the old monthly round-up it still has a very special, new, different, accurate guideline, especially for 2019. Ganesha says Bejan is now eighty-seven. It is his voice of experience which says it specially for the year 2019. Let me explain to you that astrology is mainly about the right person being at the right place and at the right time and therefore this blueprint for the year will be of special importance and meaning to you.

Ganesha asks for a look at the decans. Ganesha says as you may know, each sign occupies 30 degrees of the zodiac. These can be divided into three decans of 10 degrees each. While everyone born between 23 January and 22 February is an Aquarian, the traits will vary according to the decans.

If you were born between 23 and 30 January (the first decan), you are ruled by both Saturn and Uranus. These endow you with the persistence to overcome all obstacles, and also with originality and creativity. You will naturally do very well in the world with this practically unbeatable combination.

If you were born between 31 January and 9 February (the second decan), your ruling planet is Mercury, so that you have the capacity to reach out to people, along with an ever-ready wit and fine reasoning ability. You really consider variety to be the spice of life and are changeable by temperament, though you are also mentally compatible with most people.

If your birthday is between 10 and 22 February (the third decan), Venus is the ruling planet, making you attractive to the opposite sex, and romantic, but this may lead to much heartburn and misunderstanding. You can appreciate both the beauty and utility of any plan or enterprise. Your restlessness makes it difficult for you to relax.

Jupiter, the good-luck planet, will be in your eleventh angle. The eleventh angle shows the following:

- Funds, investment, joint finance, buying/selling/investment opportunities, real-estate deals, success in gas and chemicals and heavy industry.

- You will enlist in clubs and organizations, and socialize till you almost drop dead! You will have friends and well-wishers to help you along. That's a big blessing.

- You will receive gifts, perhaps, some inheritance, or unexpected money, or bonus, and one wish might come true. I am truly happy for you.

- Jupiter, the 'wisdom through plenty' planet, will be totally and absolutely on your side. It will certainly act as a ballast, a counterweight to the lash of Saturn. Also, it will make you confident and moneyed and a lover of the good things of life.

- The two sectors or frontiers that Jupiter will galvanize will be work and foreign connections. That's for sure, says Ganesha, and so shall it be.

- Journeys, festivities, celebrations, invitations, functions, research, travel and ties in an interlink, and pilgrimages will keep you mighty busy. Yes, you will widen the vistas of your horizon in a million different ways – say, reading, TV, having a guru, cultivating rare insights and so on. My astrological tips are, be reasonable and do not take slights and hurts too personally.

But Saturn will be in your twelfth angle or hidden angle or mysterious angle or esoteric angle from 21 December 2017 to 23 March 2020. Here I openly admit that even after eighty-seven years it is difficult for me to predict only about Saturn for Aquarians. Therefore, I am taking help of Dell Horoscope very openly and informing you about what it says about your twelfth angle. The twelfth house rules:

- personal limitations and problems,
- situations of a secretive nature,
- investigation, self-analysis,
- charitable undertakings.

You achieve the greatest success by

- gaining insight into your difficulties,
- sticking to a very regular routine,
- surmounting oversensitivity,
- strengthening faith.

You should avoid

- inviting unwise romantic entanglements,
- creating tension with your subordinates,
- over-magnifying your troubles,
- not realizing your faults.

My personal solution is that you should recite the mantra of Hanuman-ji on a Saturday. The mantra is: *'Om pram prim praum sa shanaye namah'*. Recite this mantra twenty-one times and give money very specially to lepers, lame people, old people and all orphans. Believe me Ganesha, Hanuman-ji, Shiva and Allah always love those who give charity to others. If you are charitable to others God will be generous and kind to you also. This is an eternal law which will never change.

Uranus is your main planet. Uranus will be changing signs in 2018–19. This change of signs will give you sudden money and good food from time to time. You can wear multicoloured clothes for your benefit. This is a special tip only for you, says Ganesha.

Your months of pleasure and entertainment will be April, June, October and December. You can throw a party and call friends over and have a nice time if you like it. Electronics and all modern media of information will be useful as well as pleasurable for you. You will make merry and even marry in 2019 if you wish it. That is the last word of Ganesha.

Violet is your flower and all new methods and science, and metaphysics and inventions will be up your street as we say.

Your body salt could well be calcium and magnesium.

Your happiness quota will be 81 per cent.

SPECIAL BONUS

Ganesha says my devotee Bejan is introducing the topic of special bonus for all of you. This special bonus is placed upon Mercury. Mercury represents journeys, ties, trips, mobility, migrations and all types of communication. Venus represents love, beauty, the fine arts, polish and finesse, diplomacy and persuasion, joy and a big hurrah to life itself. Therefore, I am combining both communication and joy for you and giving you the necessary dates.

Property and travel come very specially under Venus for you Aquarians. The time for it will be 7 January to 3 February, 2 to 26 March (particularly), 21 April to 14 May (things fall into praise naturally), 9 June to 3 July (money and honey), 28 July to 20 August (love and legal issues in a strange and contradictory manner), 15 September to 8 October, 2 to 25 November (on top of the world), 20 December to 13 January 2020 (pending work will be done).

Hobbies, entertainment, creativity and profit, all comes under the orbit of Mercury for you Aquarians. This is how it goes. Mercury will help you from 24 January to 9 February (many opportunities in every possible away), 17 April to 6 May, 21 May to 4 June (by your mental brilliance you win the game of life), 27 June to 18 July (money and honey), 11 to 28 August (trips and ties), 14 September to 2 October (many opportunities to be happy and helpful), 9 to 28 December (friends and well-wishers and the boss will help you).

However, I must be true to my own self. This is only a solar-scopic reading based mainly upon the position of Sun. Therefore, it may not be as accurate as a personal horoscope. At the same time, I have put in a lot of effort, imagination and inspiration as well as intuition into it. In other words, I have tried my best. The results are up to our maker. Yes, sometimes miracles do happen. Therefore, I end on a positive note, though I know my limitations very well.

KEY TO WEALTH BY NASTUR DARUWALLA

Aquarians are intuitive, independent and stubborn. They hate anything incomplete. The mantra for them is: *'Aim hrim shrim ashtalakshmyai hrim hrim siddhaye mam grihe agachchha namah swaha'*. They must give food to the needy.

WEEKLY FORECAST BY PHASES OF THE MOON

6 January: New Moon in Capricorn

You start the year by spending heavily on home and family. There is a possibility of foreign collaborations and links, both personal and for work. You may need to make a decision about a trip, or a shift to another city or just moving house. At the same time, there are genuine psychic impulses coupled with a mood of introspection.

14 January: Moon's First Quarter in Aries

Once again, I am compelled to repeat (not ad nauseam, I hope) two of my favourite phrases. One is of course, 'trips and ties', and the other 'reaching out to people and places'. Both these, used together, just about sum up this quarter of the Moon. Signing of deeds/documents, public relations, publicity, sales, research and, on the personal front, relationships from engagement and marriage to separation and divorce are covered.

21 January: Full Moon in Leo

Much of what has been written in the forecast just above will apply, only more so in this quarter of the Moon. In fact, all in all, it is the clincher, the culmination in terms of work, promotion and efficiency, and of course, the good results and rewards that go with them. You will think big, perhaps even think global (in all meanings of the phrase), act with foresight and intelligence. Those in management will shine particularly.

27 January: Moon's Last Quarter in Scorpio

Your personality too will extend itself to the world now, not just your work and management skills. In fact, January itself as a month is a precursor to all that will befall you this whole year. Uranus will remain in your sign till 10 March 2023, and will continue to influence your growth and development. You may be jerked out of your complacency, and grow into maturity.

4 February: New Moon in Aquarius

This phase of the Moon may cause tensions and some health problems. You may have to give up something to get something better. Your major concerns will be the home, your marriage, attachments, along with journey and, of course, money, which is important, but not necessarily scarce.

12 February: Moon's First Quarter in Taurus

The focus this quarter is home/house/property/land/ renovation/overhauling or installation of machinery. Parents and in-laws too will require time and attention. Investments/ trade/loans, on the money front, will be important, also perhaps a home away from home.

19 February: Full Moon in Scorpio

You will be a prey to sudden attractions or equally strong dislikes. You will take important decisions now that will influence or even radically alter the course of your life. A journey with a stopover is likely. In this intellectually and emotionally charged week, you will realign your values, reassess your friends, reappraise your enemies.

26 February: Moon's Last Quarter in Sagittarius

A time of being very sociable (not your strong point, Aquarians) and group activities, teamwork, a touch of romance and dalliance keep you busy. You will be open to new ideas and influences, and guests, relatives, even foreign visitors will see a good deal of you.

6 March: New Moon in Pisces

This phase of the Moon will favour loans and funds and, on the other hand, religious activities and ceremonies. A time to exercise great moderation in both behaviour and food, warns Ganesha. It's never good to be rash and rude or to overindulge in food and drink, even otherwise. More so this quarter.

14 March: Moon's First Quarter in Gemini

A quarter full of excitements in terms of immense creativity, the awakening of romance in your heart, and a growing leaning towards tantra, mantra, spiritualism. Travel is almost definite. All in all, your personality acquires new facets and becomes more rounded, more complete.

21 March: Full Moon in Libra

The strongest trend will be that of money, laying financial foundations. Buying and selling are very possible. In a short while, maybe just at the end of the month, new prospects will open, both for recreation and enjoyment, and for work.

28 March: Moon's Last Quarter in Capricorn

There's no looking back for you from 29 March, with the favourable placing of Mars and Mercury – which gives you power and energy, and inspiration, respectively. The theme of new avenues of earning and new prospects becomes much firmer. An exceptionally favourable phase says Ganesha.

5 April: New Moon in Aries

This quarter is for partnerships at all levels ranging from purely personal to purely professional/business or a combination of both. You may be manoeuvring/jostling for power and position. Ganesha advises a more laid-back approach, not too much push, and not too much aggression.

12 April: Moon's First Quarter in Cancer

Things will certainly start looking up now. Entertaining, entertainment, travel and trips, contacting and interacting with people who matter, loving and being loved – are all favoured now. Also, you receive joyful news, will be neighbourly, and serve the community, organize loans. Lots of action.

18 April: Full Moon in Libra

A time when you will enjoy fully all the pleasure that the senses can give. A fine time for artists and painters, connoisseurs, gourmets, all those concerned with the art of good living. I must ask here, is there any other way to live? You will also have much to do with family values and financial transactions. A wonderful trend for the month, says Ganesha.

26 April: Moon's Last Quarter in Aquarius

Action, work and employment are the focus. In fact, those not employed may get jobs; or better prospects, for those who do not have suitable jobs, are favoured. It's the right time to develop a good work ethos, because the rewards will come. Ganesha guarantees that.

4 May: New Moon in Taurus

A quarter of immense satisfaction for you, both for work and for relationships at all levels, even for recreation and play. Family, work, entertainment, children and hobbies, creative pursuits – all give pleasure, especially the satisfaction of a duty well done.

12 May: Moon's First Quarter in Leo

This previous quarter made you ready to take on the world, chance your arm! You will now be romantic, even ardent, full of passion and love, and enthusiasm. You will have the drive and determination to take on anything you want to try your luck. Travel is particularly important, along with meeting people from distant lands, relatives, tourists, visitors. You will extend your hospitality too.

18 May: Full Moon in Scorpio

Financially, an upswing this quarter. Loans and funds will materialize for work projects. Investments and capital formation will be expedited, and so also finance for new projects. Employment prospects will also improve for those Aquarians who are already employed. A house move/shift or even immigration is favoured.

26 May: Moon's Last Quarter in Pisces

This is the real time for trips, relationships, loans and finances, to work out smoothly. Conversely though, it's also the time for coming together or separating permanently. That's the way the stars work for you, says Ganesha.

3 June: New Moon in Gemini

News and views will take top priority. It is good news for you and bodes well for the future. It is also the time for mutual giving and taking in relationships, for interfacing with as many people as possible.

10 June: Moon's First Quarter in Virgo

Once again, the focus is on finance, but it's through the opposite of a zoom lens – it's long-range, of years to come. I mean by this that your activities now will create avenues for making money in the long-term future, not the immediate future. You will be building up your financial muscle and clout. Till the middle of June, house/home/family will also demand attention.

17 June: Full Moon in Sagittarius

This quarter falls in your eighth sign, as per Western astrology, emphasizing the personal angle totally. It is a time of warmth, intimacy, caring and sharing with family, parents, kin, in-laws. A clan gathering or family get-together may occur. Religious rites, some type of reverence paid to ancestors may also be performed or participated in. It's the household gods and Ganesha, of course, for you this quarter.

25 June: Moon's Last Quarter in Aries

Work may well become the centre of your existence this quarter, as it will overshadow almost all other concerns. You may receive secret help and guidance in your endeavours. However, it is a slightly tricky time with regard to health, so a little care and some safeguards may be necessary. Aquarians are supposed to be careful and meticulous. I think it's time to live up to your sign.

2 July: New Moon in Cancer

The house/home and family are in focus this quarter, and as a direct opposite, travel as well. You will also alternate between periods of rest and intense activity, bordering on manic. Home and office will come together in some way, or you could be operating from home. Also, your artistic ability will be at full flow, and greatly appreciated by others.

9 July: Moon's First Quarter in Libra

This could be the time not just to make decisions but to execute them, especially where partnerships, business or marriage and even migrating to foreign lands are concerned. Personal issues like health, children, a major house move are important too.

18 July: Full Moon in Capricorn

Finances will improve, in the sense that buying/selling/trade/ representation could be important and successful. You could even start up a new venture. The theme is opportunities for business – tenders, contracts, hypothecation, loans can all go in your favour. Taxes may need to be paid and expenses met.

25 July: Moon's Last Quarter in Taurus

Entertaining for business and pleasure is more than likely. Also, an increase in earned income, but you have to slog to earn it. Your inner world, the real you, are likely to matter a lot to you. In your quest for peace and happiness, you will learn (if you haven't already done so) to make compromises.

1 August: New Moon in Leo

Personal issues are once again focused, along with funds and finances, a yearning for emotional stability and peace. In that sense, all the trends of July converge now. Acquiring a home/property/vehicle will all be favoured from now till the end of August.

7 August: Moon's First Quarter in Scorpio

Socializing, group activities, hospitality will all win you great appreciation. Money will flow in freely, generally making life much easier. A wedding or engagement may occur too. Luck is on your side in several ways now.

15 August: Full Moon in Aquarius

This phase of the Moon will make you have a lot to do with work and also with projects, dependents including pets, even colleagues. You will have a strong sense of social service. But you'll be working very hard indeed to cope with all this, in your steady way.

30 August: New Moon in Virgo

Weddings and engagements are strongly favoured, so also business partnerships, alliances, collaborations. Once again, lots of money but also lots of activity and lots of work. Keep going, says Ganesha.

6 September: Moon's First Quarter in Sagittarius

A very lucky day for Aquarians – 6 September. This quarter will prove to be a successful launching pad for travel and communication, love and legal hassles. Acquiring property, vehicles and/or assets is favoured. You will also acquire a fresh new outlook and perspective on life.

14 September: Full Moon in Pisces

The impact of the previous phase of the Moon continues. Your work and efforts could show results; promotions, perks and recognition could materialize. You will feel content with your lot. Spiritualism and religion will also hold a special interest for you.

22 September: Moon's Last Quarter in Gemini

This phase of the Moon grants wish-fulfilment, success in a recent venture or enterprise. You can be sanguine even, not just confident. Publicity, advertising, sales and communications are all favoured too. It's the right time to get things done. Deeds and documents that you start to process now will be signed by November.

28 September: New Moon in Libra

The angle of finance, family and journeys is favoured and will work for you. There will be opportunities to travel for business or pleasure, acquire skills and learning, and of course, to make money too. Another fine quarter as a gift from Ganesha.

5 October: Moon's First Quarter in Capricorn

Finances at all levels are stressed, and also health. Lotteries and sudden gain are likely, but health may be suspect. You have been working like a beaver and may have to pay the price. Rest and adequate care are what Ganesha prescribes.

13 October: Full Moon in Aries

Alliances and contacts, home and family, marriage and divorce are all in focus. Therefore, one can safely say that relationships are all important in this quarter. A journey with a stopover is also likely.

21 October: Moon's Last Quarter in Cancer

It's time now to push with all you've got. It's time to exercise your skills at wheeling and dealing, diplomacy, even be devious and crafty if the occasion demands it. You will certainly need to be a tough negotiator to outwit your enemies/rivals who are waiting for you to slip up. The thrust of all this will naturally be money.

28 October: New Moon in Scorpio

You will experience a release from tensions, relief from care, under the influence of Venus. Your work, actions, emotions will all benefit. Better health too, and more rest. You will be able to deal better with added responsibilities and more fun and games, more laughter, more love. PR exercises, meets, conferences, seminars will be handled with confidence and panache.

4 November: Moon's First Quarter in Aquarius

Courage and resourcefulness will be your middle names this quarter. Ambition to the extent of being an obsession will be yours too. You must take care not to let it degenerate into mere aggression. Ganesha advises you to soften up a little, be less pushy or less demanding, especially with family and loved ones, subordinates and colleagues. Parents and elderly relatives have an important role this quarter.

12 November: Full Moon in Taurus

A pole vault in your career/business/profession is almost definite. What's to be controlled is how you handle it. Once again, a tendency towards aggression has to be curbed. Soft-pedalling, particularly with children, will stand you in good stead, particularly in November.

18 November: Moon's Last Quarter in Leo

You will be on the fast track. Wish-fulfilment and truly getting places are foretold. Also, the birth or conception of a child is a definite possibility (I may be wrong there – the ultimate decision is yours!) New ventures will be successfully launched, existing ones will bring gains. Opposition to plans will dwindle away.

26 November: New Moon in Sagittarius

Matters of the spirit, the other world, of higher learning and metaphysics, even tantra and mantra, if so inclined, will draw you. Also, a good time for innovations, inventions, electronics and supercomputers. You may take on extra work. Journeys and travel are a result of work, or could be triggered off by your inner restlessness.

4 December: Moon's First Quarter in Pisces

Once again you are searching within for spiritual riches even as you go all out to cope with and win in the material world.

Communication will be the buzzword at this point. Anything to do with house/home right from buying to redoing will claim your attention. Supernatural forces seem to control you at this time.

14 December: Full Moon in Gemini

The inner world is reconciled with the outer now. You seem to be receiving divine guidance as you make the right moves, get things done, handle people with finesse and sympathy. Your status and prestige will register an increase.

19 December: Moon's Last Quarter in Virgo

Management will be your forte now – inventing new techniques and skills, perhaps even organizing meets and functions. You will be more productive, more hard-working, with infinitely better health in this quarter. Sport outings, hobbies and children – all keep you busy but happily so. Ties, bonds, relationships and marriage are all favoured too. A good phase, comments Ganesha.

26 December: New Moon in Capricorn

You will learn to manoeuvre and negotiate to get to the heap, even use secret contacts and connections. It's a good time for journeys and collaboration, and surprisingly, also for conspiracies, behind-the-scene activities and string pulling. Projects started at the beginning of the year are happily concluded and personal affairs come full circle.

PISCES

23 February–20 March

Pisces is the sign of the Dreamer-Poet. You Pisceans are psychic, often moody, hypersensitive, impractical, secretive, kind, self-sacrificing, drifting, escapist, compassionate, gentle. Ganesha points out that the three greats of all times – Michelangelo, probably the greatest artist ever, Steve Jobs, the computer wizard, and Albert Einstein, the mightiest scientist ever – were all Pisceans. This clearly proves the insight, inspiration, imagination, inventive genius, incisiveness and the creativity of the Pisceans. I will give full marks to you Pisceans and remember I am not a Piscean and therefore I can be completely objective about it. In short you Pisceans can save not only yourselves but also the whole world. Great compliment well deserved.

This time I am following it up with your digital plan for the year so that you can focus very well on it in advance and thus prepare yourself by making a blueprint of the entire year. The digital plan for the year will be short and to the point, and though it is a general reading, it should apply quite a bit to you.

BLUEPRINT FOR THE ENTIRE YEAR

January: A golden harvest for the trouble taken and the seeds planted, and that says it all.

February: Expenses, work, contracts, secret affairs, heart illuminations, though there could be inflammation of your (foot) sole.

March: Wishes granted, rewards realized, wish-fulfilment is possible; you will feel wonderful and strong, ready to take on all comers.

April: Finance, food, family, and that does mean entertainment, amusement, doing the social rounds.

May: Gains, friends, children, creativity, group activity, joy and delight in life.

June: House, home, parents, in-laws, a home away from home, travel, get-togethers and separations.

July: 'Open sesame' to fame, fortune, children, romance, hobbies, creativity.

August: Health, work, colleagues, irritations over pets, projects, and other trifles.

September: Collaborations, partnerships at all levels, a journey with a stopover, reaching out to people, places.

October: Joint finances, insurance, loans, public trusts, low vitality, sex and love in strange mix.

November: The luck of the draw, knowledge, evolution, wisdom, ancestors and rites, genuine spirituality, long-distance connections, pilgrimages.

December: A high-powered month for work and play, prestige and promotion, parents, in-laws, boss and life-mate.

Ganesha says this is certainly the old monthly round-up but specially for 2019 this monthly round-up becomes very necessary and precise and proper because of the planets Jupiter, Saturn and Pluto helping you in every possible way. Therefore, though it is the old monthly round-up it still has a very special, new, different and accurate guideline, especially for 2019. Ganesha says Bejan is now eighty-seven. It is his voice of experience which says it specially for the year 2019. Let me explain to you that astrology is mainly about the right person being at the right place and at the right time and therefore this blueprint of the year will be of special importance and meaning to you.

Ganesha asks for a look at the decans. Ganesha says since each sign has 30 degrees, there are three decans in it, which are divisions of 10 degrees each. Though all born between 23 February to 20 March come under the Sun sign of Pisces, we can know more about the individual personality traits from the decans.

If you were born between 23 to 28 or 29 February (the first decan), you are ruled by the twin planets Jupiter and Neptune. Though it is a decan of power, it also denotes fixity of purpose and obstinacy which may sometimes be detrimental to you. It also makes you difficult to ignore. Fortune favours you in middle age. Your life is eventful and full of sudden upsets. You also have spiritual and visionary qualities.

If your birthday is between 1 and 10 March (the second decan), you are ruled by the Moon. You have tremendous inner strength, but change is your middle name, especially if you are a woman. You are artistic, psychic, a healer, good at import and export. You cherish your home but lack stability and steadiness. Be careful to avoid drugs and liquor.

If you were born between 11 and 20 March (the third decan), Mars is your ruler, giving you much force and a streak of practicality. You seek knowledge as well as wisdom, but are still a Piscean dreamer. You have an unfortunate tendency to haste and irritability, which you must curb to avoid setbacks. You are interested in art, adventure, films, magazines and business enterprises.

Your major planet which has a big say and influence is Jupiter. For you, Jupiter stands for utility. Jupiter will be in your tenth angle from 9 March 2018 to 2 December 2019. The tenth angle stands for the *highest point*, the *apex*, the mid-heaven in the horoscope. It shows action and achievement. Jupiter in this tenth angle gives a major thrust, a big boost to your actions and achievements.

I openly admit that in this book I use mainly Western astrology. But Ganesha wants me to use Vedic astrology very specially for you. Therefore, these are my predictions. The tenth house also refers to popularity, authority, responsibility, the three personality factors which decide our destiny. Parents and in-laws and our duties to them also characterize the tenth house. Business and profession, but not employment, come under the orbit of the tenth house. But please remember not to rely overmuch on others, though I know you are often tempted to do so. Very often you may find that you have to travel to get your work done. Collaborations, both local and foreign, are a special feature of Jupiter in the tenth angle. If you are into publishing, the media, medicine, astrology, law, foreign affairs, travel, higher education, international affairs, so much the better for you. You will shine bright and splendidly. Perhaps the most important legacy or result of the Jupiter in the tenth angle is to make you positive, confident, ambitious.

The tenth angle stands for the apex, the highest point, the harvest, the summit of success, rewards, recognition, accolades. It is the garland the world has given you in recognition of your services. In addition, the tenth angle stands for:

- profession, function, source of livelihood, government service, honour from the government, business;
- status;
- wealth;
- political power;
- progress;
- hidden treasure;
- dominance;
- proficiency;
- father's wealth;
- well-being;
- foreign travel;
- place of residence;
- performance of secret and religious deeds.

Now we take Saturn according to the great astrologer Robert Hand. Saturn will be in your eleventh house from 21 December 2017 to 23 March 2020. The eleventh house represents friendship. This can refer to your feelings of friendship for each other as well as to friends that you have as a couple.

In the first instance, Saturn in this house does not make friendship between the two of you impossible, but it does somewhat cool your expression of it; you may still have a very enduring relationship. You may be quite reserved with each other, but that may be just as well, for it will permit your relationship to last longer. At the eleventh house Saturn can indicate a very long-lasting friendship.

The eleventh house is also the house of one's ideals, hopes, and wishes. In some cases, Saturn can signify that the two of you have very different ideals and that you don't usually react in the same way towards things. This can diminish your sympathy for each other and make the relationship quite difficult. Taking the second side of the eleventh house, that of friends outside the relationship, it can be said that when Saturn is operating positively, it indicates few outside friends, but those few will be firm and long-lasting. They may be older than either of you. But if Saturn's influence is not working out well, you may not have any outside friends, probably because of some rigidity in you that makes it impossible for you to share yourselves with others.

Let me add a word or two to the prediction of the mighty Robert Hand. In 2019, Saturn, Neptune and Uranus will be in a fine position to each other. In simple language, it means power, progress, prosperity and promotion for you.

Your months of pleasure and entertainment will be May, July, September and certainly November. We all live at least once; that means in this time there is a possibility of reincarnation and being born again and again but nobody can verify it completely. So my humble suggestion to you is to live your life fully. In other words, live your life to the hilt and have a rollicking good time. The choice is yours.

Your happiness quota will be 86 per cent.

THE LAST WORD

Today Tuesday, 9 January 2018 is the day of Ganesha when I am writing this. Therefore, in the name of Ganesha, I have to explain to you, good people, that not by astrology but by astronomy Jupiter is the greatest planet. Jupiter is the saviour and it has a special role over the signs Pisces and Sagittarius.

SPECIAL BONUS

Ganesha says my devotee Bejan is introducing the topic pf special bonus for all of you. This special bonus is placed upon Mercury. Mercury represents journeys, ties, trips, mobility, migrations and all types of communication. Venus represents love, beauty, the fine arts, polish and finesse, diplomacy and persuasion, joy and a big hurrah to life itself. Therefore, I am combining both communication and joy for you and giving you the necessary dates.

For you Pisceans, in particular, Mercury represents marriage, collaborations, ties, lawsuits and journeys. Therefore, Mercury is mighty important. This is how it goes. Mercury helps you from 5 to 23 January, 10 February to 10 April (excellent in many different ways), 17 to 20 May (work will be done), 5 to 26 June (by your charm and wit you will win the game), 19 July to 10 August (a second chance to be happy in life), 14 September to 2 October (the turn of the dice of life will be in your favour) and 9 to 28 December (you will be the top gun in your field of activity).

For contacts and connections Venus will be essential and lucky. The period for it will be 1 to 6 January, 4 February to 1 March, 27 March to 20 April, 15 May to 8 June (very particularly), 4 to 27 July (outstanding), 21 August to 14 September (contacts and contracts), 3 October to 1

November (give your best shot in whatever you do) and 25 November to 19 December (boss and well-wishers will help you).

However, I must be true to my own self. This is only a solar-scopic reading based mainly upon the position of Sun. Therefore, it may not be as accurate as a personal horoscope. At the same time, I have put in a lot of effort, imagination and inspiration as well as intuition into it. In other words, I have tried my best. The results are up to our maker. Yes, sometimes miracles do happen. Therefore, I end on a positive note, though I know my limitations very well.

KEY TO WEALTH BY NASTUR DARUWALLA

Pisceans are sensitive and giving. They do things with patience, care and love. For them the mantra is: *'Om shrim hrim shrim kamale kamalalaye praseeda praseeda; om shrim hrim shrim mahalakshmyai namah'*. Water is their main element, so give pure drinking water or water purified by reverse osmosis to the needy and the birds.

WEEKLY FORECAST BY PHASES OF THE MOON

6 January: New Moon in Capricorn

This phase of the Moon makes you highly creative, emotional and sensitive to your environment and loved ones. Children, art, writing, film-making, research, even domestic chores and, interior decoration – you do excel at whatever you turn to. You are in the driver's seat from the word 'go'.

14 January: Moon's First Quarter in Aries

Finances are now spotlighted. Loans, funds, pensions, securities, joint finances, ex-gratia payments and bonus are favoured. You now have the financial standing to make the long-desired changes. You can thus make important decisions.

21 January: Full Moon in Leo

The supreme point or culmination comes for you in this phase of the Moon. Relief from tension and opposition, and romance and glamour return to your life. More money and good investment opportunities will come your way. Also, journeys for both pleasure and gain. This monthly trend will be the forerunner to happiness, and to better things in life.

27 January: Moon's Last Quarter in Scorpio

Power, money, good fortune and vision and sensitivity – right now you enjoy them all. You will display shrewdness and practical knowledge, class and intelligence in all your activities and the results are bound to come, Ganesha assures you.

4 February: New Moon in Aquarius

Pending work will be completed and something new will be started. A whole range of possibilities is covered: deeds and documents, visitors and relationships. A wish-fulfilment is a definite possibility. Here, the deciding factor will be the personal element.

12 February: Moon's First Quarter in Taurus

Your job, profession and business will be accentuated. You will be in a position to give and receive favours. You will gain in maturity, generosity and understanding too, and should use all these qualities in your dealings with subordinates, colleagues and boss. Health may pose some problems.

19 February: Full Moon in Scorpio

Much of what went before will be stressed, and will crystallize this quarter. You will strive for an intense stillness, if not true peace, to provide guidelines to your life. You will instinctively acquire the knack of doing the right thing at the right time, and it will be duly reflected in all your activities.

26 February: Moon's Last Quarter in Sagittarius

A period of introversion, perhaps as a natural corollary to the preceding weeks of this month. House, home, parents, buying and selling, shopping and renovation will all keep you occupied. You will thus be functioning at several levels all at once.

6 March: New Moon in Pisces

You will be extending help, advice and hospitality to people, sincerely and generously. However, a tie or relationship may be broken around this time. Material success and emotional bonding are both favoured by the full Moon, so either a new bond will be made, or you'll find the right mindset to cope.

14 March: Moon's First Quarter in Gemini

The sting of sorrow and melancholy may be experienced towards the end of this quarter, but on the whole, it's a time for healing, coming to terms with yourself, learning your karmic lesson. It may well be the end of the tunnel for you before emerging into the light, and your heart, spirit and relationships will all benefit.

21 March: Full Moon in Libra

All your talent traits and strengths will emerge. You will experience almost a dramatic rebirth. New beginnings are foretold by Ganesha. It is a time of preparation, of gearing up for April. Society, your surroundings, nature and God, you own self, will all be in fine tune with each other.

28 March: Moon's Last Quarter in Capricorn

You will be bouncing back with a vengeance now. The personal and professional aspects will now converge and meet. You will have the energy and ability to get on with life and living. The hard time you've had finally comes to an end now. You will find life an adventure.

5 April: New Moon in Aries

The time for new work and new relationships, and also the end of an enterprise. An engagement, marriage or a divorce may occur. But you will be dexterous and versatile in your ability to cope.

12 April: Moon's First Quarter in Cancer

You will finally have an easier time. Practical matters will come to the forefront. There may be temporary cash-flow problems but never such that you cannot cope. You will be pragmatic and level-headed in your approach to work, profession or business.

18 April: Full Moon in Libra

This phase of the Moon grants you enjoyment of all the pleasures of the senses. Family values and financial transactions will both be important. Artists, painters, connoisseurs and critics will be helped, in fact all those who like to enjoy life. And who doesn't, asks Ganesha.

26 April: Moon's Last Quarter in Aquarius

Employment is the name of the game this quarter. Hard work is necessary, but the just rewards will also come, and that's a wonderful incentive. Along with job satisfaction will be the satisfaction of doing your duty and making others happy.

4 May: New Moon in Taurus

The full Moon influences your emotions and passions, making you ardent and romantic. You will be brimming with vigour and confidence, ready to take on the world. Travel, meetings with foreigners, tourists and even relatives are likely. You could even be playing host.

12 May: Moon's First Quarter in Leo

You will be inventive and ingenious but also restless and introspective. Opportunities to make money will arise. Distant places may play an important part. You will cover a lot of territory and people. Spiritualism can prove to be both a strong force and a strong interest in your life.

18 May: Full Moon in Scorpio

It's the right time, now and finally, for making a lot of headway through communication, contacts, study and research, and also travel. A readiness to take chances will yield results, but Ganesha warns you not to get trigger-happy and careless. All the trends of the preceding weeks are in sharp focus now.

26 May: Moon's Last Quarter in Pisces

You will carry on the good work and complete pending assignments. Pisceans can be inspired to bouts of brilliant activity. The trick is to sustain it. Some projects taken up now may, of course, take years to complete, but in June, you will be able to get things moving. Both adventure and good luck are foretold by Ganesha.

3 June: New Moon in Gemini

You will be dynamic and exceptionally ambitious. Money will flow in and promotion or a job switch is probable. You will need to weigh up pros and cons because ultimately the final decision is yours, not the astrologer's, advises Ganesha sagely.

10 June: Moon's First Quarter in Virgo

This quarter falls in your own sign according to Western astrology. It will make the whole month a time of warmth and personal bonding. Especially so with family, relatives, parents and in-laws. Family gatherings, homage to ancestors and religious rites are all likely.

17 June: Full Moon in Sagittarius

This phase of the Moon accelerates and strengthens all the trends foretold for the last few weeks. The impact of all this will be experienced till the end of July, in fact early August too. The emphasis is strongly on travel, communication and home, all three in practically equal measure, under the combined influence of Mars and Venus.

25 June: Moon's Last Quarter in Aries

Care and tending of the home will be in strong focus both this week and the next. Right from buying or acquiring to redoing and painting, even a house-warming party, all come within this bracket. In fact, your office may be a second home, or you may also get a home away from home.

2 July: New Moon in Cancer

You will continue to devote considerable time and energy to your home/office. They may, in fact, be combined as I've already indicated in the previous forecast. Also, your relationships with people will be important, along with chances of making some more money.

9 July: Moon's First Quarter in Libra

Finances are in focus – particularly in buying, selling, brokerage and trade-related activities. Also, loans, contracts, tenders. But expenses, perhaps taxes, will be rather heavy. This may also be because of a fair amount of entertaining.

18 July: Full Moon in Capricorn

You will be suffused and overpowered with love, also patriotic love for your country. You will be going around with your head in the clouds, full of fantasies and sentimentality. Pisceans are known to be dreamy to the point of being reckless, so it's good to keep your feet firmly on the ground.

25 July: Moon's Last Quarter in Taurus

You will be extremely work-oriented this quarter. If in government service or in the corporate world, there are great chances of a move up the ladder. Health will, however, pose a few problems, now and in the coming weeks too. Ganesha advises you that since Pisceans are often delicate and also neglectful, *care is a must*.

1 August: New Moon in Leo

You will extend yourself in many directions – primarily work but also wining and dining, entertaining and interfacing with people. You will display much charm and sensitivity in all you say and do. Once again, I must point out that health care is a must. Pisces is a sign that does not provide much stamina.

7 August: Moon's First Quarter in Scorpio

Mars and Venus and the moon connect beautifully to endow you with money and force and the comforts of home. It's an ideal time to be outgoing and extroverted, pleasing others as well as yourself. Those connected with art, or earning by it will prosper. Work satisfaction will be coupled with good money. It sounds a good deal, doesn't it?

15 August: Full Moon in Aquarius

You will have much to do with dependents (even pets and servants), colleagues, projects and the work front in general. This is a definite trend for the month. However, work will be coupled with plenty of fun, so you will work and play equally successfully. It's kind of Ganesha, isn't it?

30 August: New Moon in Virgo

You will be in emotional treadmill. Journeys and contacts are important, along with collaborations, import and export, higher studies abroad or negotiations with foreigners. A wedding, engagement or love affair is also likely. A notable quarter for the multitude of possibilities it offers. I forgot to add social service and care of others to the list!

6 September: Moon's First Quarter in Sagittarius

This phase of the Moon energizes the human relationships angle. The trend for engagement/marriage becomes stronger now. Also, much more interfacing with the public, and you could gain much from this, if you do it with sincerity and caring.

14 September: Full Moon in Pisces

Buying, selling, trade, brokerage will give you much gain. Extra attention will be required for the home/office/shop. Also likely is the acquiring of a vehicle, or comfort and luxury good that make for better living. You may even visit auctions and flea markets to pick up attractive bargains.

22 September: Moon's Last Quarter in Gemini

This phase of the Moon brings the world to your heart. Unions and separations will occur, and you may become a newsworthy person yourself. People from distant lands will play an important role in your life. Films and television may have openings for you. You have to cash in on the possibilities that lie before you once again.

28 September: New Moon in Libra

You will make that extra effort to get fame, recognition, glory. It's going to be hard work (it's never easy, believe me!). But you'll get there in the end. Ganesha advises you, in modern American parlance, to step on the gas.

5 October: Moon's First Quarter in Capricorn

It's a time to be alert and active enough to snatch the opportunities for advancement that come your way. Recognition and rewards, both here and abroad, are yours for the taking, and take them you will, in your present frame of mind. Finances and wealth will be important.

13 October: Full Moon in Aries

Once more, I must mention finances and health as being all-important. In that respect the last quarter was a prelude to this. Business funds, loans and financial settlements, perhaps even lotteries and windfalls, are all covered. You will turn to spiritualism to gain strength of mind. Your physical strength and stamina have to be carefully guarded.

21 October: Moon's Last Quarter in Cancer

This quarter helps you to 'win friends and influence people'. Sexual encounters and intimacy too are highlighted. You will have a natural magnetism and flair. Finances will be the centre of your professional existence, and remember, all this is a monthly trend.

28 October: New Moon in Scorpio

Children, care of home/office, family get-togethers, entertaining, amusements – all keep you very busy indeed. Also, likely installation of labour-saving gadgets, and acquiring luxury goods – generally making your life pleasurable and comfortable. Marriage plans, 'live-in' arrangements, prestige and power are all part of it and are strongly favoured too. 'Material gain' just about sums it up.

4 November: Moon's First Quarter in Aquarius

This phase of the Moon will make you fun-loving, daring and adventurous and very mobile and volatile. And what's more, this trend will last till the end of the year. While you're riding the wave of popularity and success, practicalities will resurface. Strictly speaking, that's true for next quarter's forecast.

12 November: Full Moon in Taurus

A crucial time for loans, funds, capital formation, investments – the entire gamut of financial matters. Equally crucial will be tantra, mantra, yoga, meditation and spiritual and occult practices. A house move may also be mooted. However, it's ideally a time for people and places, news and views too, so it's going to be truly hectic time for you.

18 November: Moon's Last Quarter in Leo

You get 'ready for action', here meaning consistent hard work which will pay dividends. You will be appreciated and will have both an excellent managerial outlook and power. All this will be a tremendous confidence booster, which is a boon from Ganesha in a month of giving and carrying out orders.

26 November: New Moon in Sagittarius

You will now hit the big times and will be occupied with a number of things and events. Your activities get a major boost with the combined influence of Saturn and Uranus. Journeys, ceremonies, marriage/engagement and publicity are all favoured.

4 December: Moon's First Quarter in Pisces

This is a quarter full of paradoxes. While everything concerning house/home (renovation, buying/selling specially) will have first place, you will be equally and heavily involved with communication and correspondence, and yet you will be drawn to spiritualism, to seeking the kingdom within. You may find yourself receiving timely divine guidance too. A truly phenomenal quarter.

14 December: Full Moon in Gemini

Matters of the spirit will continue to attract and fascinate you. Equally fascinating for you will be computers, electronics and automation. Your restlessness may take you on a journey, or perhaps make you take on extra work. Ganesha here cautions you not to overextend yourself.

19 December: Moon's Last Quarter in Virgo

Personal matters like children, marriage, the handling of meets and functions, are all focused. Conferences may have to be handled. It's here that you will display both style and confidence and also work tremendously hard, secure in the knowledge that you will get the rewards due to you.

26 December: New Moon in Capricorn

This quarter in your sign endows you with added confidence and happiness. You may have been having some health problems which should now ease considerably, with a lessening of the pressures on you.

The Message of the Zodiac

At the outset, let me state that this is not an original interpretation from me, or even a concoction. This is the way most astrologers perceive the signs of the Zodiac, and their secret motivations.

Aries:	I am
Taurus:	I have
Gemini:	I think
Cancer:	I feel
Leo:	I will
Virgo:	I analyse
Libra:	I balance
Scorpio:	I desire
Sagittarius:	I perceive
Capricorn:	I use
Aquarius:	I know
Pisces:	I believe

The 'Mostest' of Everything

Here's a fun mind game, the 'mostest' of everything by astrology. Ganesha laughs. Be ready!

Money: Leo, Scorpio, Capricorn, Taurus and Cancer are the *richest*.

Intelligence: Gemini and Aquarius are the ones with the brain cells.

Sex: Taurus, Leo, Sagittarius and Scorpio are the strongest with the maximum stamina, and sexual.

Food: Cancer and Taurus are the foodies, the gourmets (connoisseurs of good food) and gourmands (gluttons). Also, great chefs.

Artistic: Pisces, Libra, Leo, Taurus, Aquarius and Cancer are the most artistic.

Idealistic: Leo, Aquarius, Pisces, Virgo and Cancer are the most idealistic.

Sports: Sagittarius, Leo, Scorpio, Taurus, Aries, Cancer and Capricorn form the best sportspeople.

Property: Sagittarius, Pisces, Gemini, Virgo, Aquarius, Capricorn and Scorpio acquire the maximum property.

Spirituality: Gemini, Leo, Scorpio, Cancer and Pisces form the most spiritual people.

THE 'MOSTEST' OF EVERYTHING

Cars and Aeroplanes: Scorpio, Leo, Capricorn, Taurus, Aquarius, Gemini, Virgo and Sagittarius possess the maximum number of cars and aeroplanes.

Fashion: Leo, Libra, Scorpio, Taurus, Pisces, Aries and Gemini are the most fashionable.

Acting: Scorpio, Leo, Pisces, Gemini, Libra, Virgo and Sagittarius are the best actors.

Handsome/Beautiful: Leo, Libra, Scorpio, Taurus and Pisces are the most beautiful.

Rogues, Rascals: Scorpio, Leo, Gemini and Pisces form the worst in this category.

Honest: Cancer, Virgo, Taurus and Capricorn are the most honest.

Politicians and Leaders: Capricorn, Scorpio, Libra, Aries and Sagittarius constitute the most politicians and political leaders.

Gamblers: Scorpio, Leo, Sagittarius, Gemini and Aquarius are the prominent gamblers.

Bossy: Leo, Capricorn, Virgo, Taurus and Sagittarius form the maximum number in this category.

Nice and Easy: Libra, Cancer, Aquarius, Pisces and Gemini form the bulk of this category of people.

Home and Family: Cancer, Virgo and Taurus are the most devoted to home and family.

Work: Capricorn, Leo, Scorpio, Virgo and Aries form the workhorses.

Comics and Clowns: Cancer, Pisces, Gemini, Aquarius and Sagittarius are to be found in maximum numbers of comedians and clowns.

Books: Cancer, Libra, Gemini, Virgo and Pisces form the major book lovers.

Painting: Virgo, Taurus, Libra, Pisces and Gemini are the best in painting.

Industry: Aries, Taurus, Capricorn, Virgo and Aquarius are the mainstay in industries.

Singing: Libra, Taurus, Scorpio and Pisces are the best in singing.

Science and Technology: Libra, Aquarius, Sagittarius, Gemini and Aries form the main bulk of scientists and technologists.

Colours, Photography: Libra, Taurus, Aquarius, Capricorn, Pisces, maybe Virgo too are the best in photography and in matters to do with colours.

Sales/Publicity: Leo, Libra, Gemini, Cancer and Sagittarius form the best salespersons and experts in publicity.

Business: Libra, Scorpio, Capricorn, Taurus and Aries form the most successful businesspeople.

Luck: Nobody really knows, but it is believed, rightly or wrongly, that Leo and Sagittarius could well be the luckiest.

Management: Capricorn, Virgo, Taurus, Scorpio and Leo are the best in management.

Relationship: Libra, Cancer, Leo and Aries are the best in making and maintaining relationships.

Pets: Libra, Scorpio, Virgo, Sagittarius, Pisces and Gemini are mostly fond of pets.

Human beings are not simple and easy to understand. They are highly complicated. The same person can be many different things to many people. Therefore, your Ganesha devotee wants to make it absolutely clear that the above analysis may not be true at all times. Also, strictly speaking, the complete horoscope must be examined. This is only fun and games reading, which may be partly true. Enjoy! Cheat if you like! It's all a part of the game! But, be happy!

Special Note: In India we have Taurean Tendulkar in cricket; Sagittarian Dilip Kumar, Libran Amitabh Bachchan, Scorpio Shah Rukh Khan in acting; Scorpio Premjee, Capricornian Dhirubhai Ambani in finance; Arian Ravi Shankar in music; Virgo M.F. Husain in Arts; Leo Shri Aurobindo in spirituality; Cancerian Dom Moraes and Taurean Rabindranath Tagore in poetry and literature.

Gems for the Zodiac Signs

In Indian astrology, there is a verse in the *Jatak Parijat* that assigns nine gems to the nine planets. These are:

- Ruby, the gem for the Lord of the Day (the Sun),
- Glowing pearl, the gemstone for the cool Moon,
- Red coral, the gem for fiery Mars,
- Emerald, the gem for the noble Mercury,
- Yellow sapphire, the gem for Jupiter, the teacher of gods,
- Diamond, the gem for Venus, the teacher of demons,
- Blue sapphire, the gem for Saturn,
- Hessonite, the gem for Rahu, and
- Cat's eye, the gemstone for Ketu.

More on Gems

According to Western astrology this, dear readers, is the final list, if you have to select only *one* gem. Yes, for Scorpio and Aries, your Ganesha devotee prefers the Coral.*

GEMS FOR THE ZODIAC SIGNS

Sign	Gem
Aries (21 March–19 April)	Coral
Taurus (20 April–20 May)	Golden topaz
Gemini (21 May–20 June)	Aquamarine
Cancer (21 June–22 July)	Pearl
Leo (23 July–22 August)	Ruby
Virgo (23 August–22 September)	Emerald
Libra (23 September–22 October)	Opal
Scorpio (23 October–22 November)	Coral
Sagittarius (23 November–22 December)	Turquoise
Capricorn (23 December–22 January)	Garnet
Aquarius (23 January–22 February)	Amethyst
Pisces (23 February–20 March)	Cat's eye

*I know that a lot has been said about other gems for each sign, but as I always say, different strokes for different folks, and we always have to see what agrees with us most. If I have to, as I said, choose one gem, this is what I would recommend. Good luck!

Personalities

Ganesha says that we are always interested in other persons. The reason is sheer curiosity. Curiosity is a sign of both relationships and intelligence. Nobody in life is perfect. We all have our faults and our strong points. We are all human beings.

Donald Trump

Ganesha says my devotee has hit hard at Donald Trump in his 2018 horoscope book and again he says, 'Many Geminis are kind-hearted, versatile and totally brilliant but Trump the Gemini is an exception.' He is a bully, insular in his thinking and behaviour, always likes to be the centre of attraction, very aggressive, and rightly or wrongly he has managed to convince people that America comes first. I may be wrong. But it is my conviction that he is certainly a bad president and will lead to confusion, complications and conflicts. I know very well that Trump and Kim Jong-un signed an international agreement on 12 June 2018. It was all about denuclearization. I hope I am wrong but frankly, I doubt how long the agreement will last. Perhaps I am prejudiced against both Trump and Kim Jong-un. But my astrological knowledge and gut feeling are certainly against both of them.

Narendra Modi

Narendra Modi is a Virgo born on the seventeenth; so also are Angela Merkel of Germany, Mrs Michelle Obama and Muhammad Ali – all born on the seventeenth of their respective birth months. I have met Narendra Modi more than once. A good organizer, a good human being, his mission in life is to purify the Ganga. My personal belief is that he is good for our country. Despite difficulties and minor hurdles, Ganesha says the BJP will again be in power after the 2019 Lok Sabha elections.

Angela Merkel

Angela Merkel of Germany is a Cancerian. She has the necessary qualities of a fine leader. But it is her sensibilities and her way of doing things including compromise at the right moment which makes her outstanding. The very fact that she has taken in so many refugees deserves praise and applause. A very worthy leader of a very great country.

Vladimir Putin

Vladimir Putin the Libran is a man of many parts, not all of them perfect and good. But he has the imagination, the organization and the past experience as head of the KGB to make an efficient leader, though not a great one. The next two to three years may find him pushed into a corner from where it will be difficult to get out.

Xi Jinping

Like Donald Trump, Xi Jinping is also a Gemini. But he is far more intelligent, practical and worldly-wise to be trapped and pushed around. He will be more than a match for Donald Trump but he is also very hard to deal with and may not get along very well with India. India should be very careful in all her dealings with China. It is a tough proposition.

Kim Jong-un

Capricornian Kim Jong-un is hard-headed, strong-willed and certainly overtly ambitious and exceptionally egoistical. In short he is a tough nut. All tough nuts should be dealt with cautiously but with infinite wisdom and diplomacy. He is not the best of leaders is all that I can say. I repeat here once again that I am not God and I could be wrong. The simple reason is I have not met any of these leaders except Narendra Modi personally.

Dalai Lama

Cancerian Dalai Lama is both practical as well as spiritual. It is a rare combination. He is good and great. He deserves praise and a powerful position in the scheme of Nature Herself. Those who harm the Dalai Lama will end up by harming their own selves. To me at least Dalai Lama is a man of goodness and Mother Nature.

Theresa May

Both Margaret Thatcher, the iron lady of England, and the present prime minister of England, Theresa May, had Libra as their Sun sign. But May, though good and powerful, is not a patch on Margaret Thatcher. Yes, I admit openly and sparely that I may be proved wrong. I am open to correction.

The Three Khans

All the three Khans, namely Shah Rukh, Aamir and Salman, were born in the year 1965. They will certainly be in the position of glory for another two to three years. Personally I find Aamir the most creative, Shah Rukh the most charming and Salman the most popular. But you, my dear readers, have a right to your own opinion.

Virat Kohli

Stylish, dashing and handsome Virat Kohli, a Scorpio, is the darling of the masses and rightfully so. He scores runs with flair and matchless grace and his cover drive has hearts throbbing and minds wildly excited. He is great. But can he outshine, surpass Tendulkar and Lara? Both are earthy Taureans. Bradman, the greatest batsman ever, was an earthy Virgo. Therefore, to me, a batsman with earthy signs could well be the greatest.

But I conclude with my own typically, humorous, sporty verse:

> In love and in cricket
> Even the great Daruwalla can lose his wicket.

Vashishth Mehta

> Oh death, where is thy sting?
> Oh grave, where is thy victory?

Yes, last year I paid my homage to my guru Vashishth Mehta in my book. Shortly afterwards he went to meet his father in heaven. This is to acknowledge his spirituality, his versatility, his greatness and his unique ability to synthesize all the planets in astrology. To me at least he was, and will always be, the greatest ever. I miss him more than ever.

Bhupendrasinh Chudasma

To the credit and grace of the education minister of Gujarat, Bhupendrasinh Chudasma, he visited me on 23 April 2018 at the Avron Hospital as I had a brain stroke and was recovering from it, and finally was miraculously saved. Chudasma shared with me a great devotion for the mighty Lord Ganesha, and I am very sure that the blessings of Lord Ganesha will make him both a good and a great man. As a person he is kind, dignified and exceptionally human.

Lionel Messi and Christiano Ronaldo

I am writing this piece on 3 July 2018 when the FIFA World Cup is in full swing. As we all know the choice is between two ace footballers, Messi and Ronaldo. I personally prefer Messi because he is more human and humane and a better human being. Messi's three-touch, technically perfect strike has proved that he employs minimalism to absolutely maximum effect, but you are of course welcome to your own opinion.

Devendra Fadnavis

Ganesha says Cancerians are intuitive, inspirational, kind-hearted, cooperative and, most certainly, very helpful to others. A fine example of it is Devendra Fadnavis, chief minister of Maharashtra. I have met him twice and therefore I speak from experience and observation. According to me, he will win the elections in Maharashtra in 2019 and continue to do plenty good. Jupiter in Libra as per Western astrology makes him dignified, balanced and wise. Saturn in Taurus makes him practical and has given him an artistic wife. Also there are only two master numbers in the world, namely eleven and twenty-two. Reason why? You cannot be born on the thirty-third or the forty-fourth or the fifty-fifth. The Moon which represents the mind is in Pisces. It shows the ability to pick and choose the right person by instinct and intuition. What comes naturally is mostly right. Therefore I believe that this man is not only good and gracious but also a natural in politics. He is made for it. He will go great guns.

Others

Neeraj Bajaj, the son of my late friend and patron Ramkrishna Bajaj, has once again proved his benevolence and fair play in dealing with the unfortunate people of the world. Dadi Mistry who had been the spokesperson for the entire Parsee community is praying for my health and well-being. Prince Lakshyaraj Singh of Mewar City Palace, Udaipur, has proved that he is both a visionary and a man of action who will do much for his country and humanity. Arvind Mothi and Anu Mothi are still waiting for my complete recovery to do a documentary on my life and times. To crown it all, the wonderfully sensitive poet Ankit Trivedi is ready to translate my book of poems into Gujarati, and Jaimin Oza will be in charge of organizing the entire affair. These are good, noble, intelligent and capable friends I am lucky to have. I salute all of them because they certainly deserve it. It is because of them that I am still alive and kicking at eighty-seven summers.

World Horoscope 2019

Ganesha says integrity is most important in life. Therefore an astrologer must take a stand even though he may be hopelessly wrong. The stand I take is that the years 2021 to 2023 is when science and technology will reign supreme and change the very fabric and direction of all of us. The possibility of space travel will change us forever.

The other important areas are: artificial intelligence; healing broken bodies, mind and spirit (most important); managerial skills of a high order; immense strides in medicine and specially brainpower; space travel; robots; biopharmaceuticals; origin of life; God vs technology; climate change; and finally the marriage of technology and humanity.

The trade wars among nations, the sheer brutality of strangulating refugees, the utter selfishness, greed and lust for power among the nations are horrendous, incredible and exceptionally shameful. But rightly or wrongly, I say that by 2021 a much better world will start to develop and evolve. Happiness is in the caring and the sharing. Baring by itself is not alone.

In the section on personalities I have said a lot and therefore I do not want to repeat. But we human beings are at our best in moments of crisis and tremendous stress. We can adopt and adapt to changing circumstances. It is possible that we human beings and robots may have a lot in common in times to come. But the future is bright, though we have to earn it. And we will earn it if we deserve it. Consequently, I am ending it on a very positive and hopeful note. We human beings are best when faced against odds. In other words at the very edge of the precipice we can show our true worth.

'Belief in belief is dead.' Poor God! He is put under a scanner and has to prove His existence. Fear is the key to life and living. Anxiety, neurosis, paranoia and terrorism, all have us by the throat. It is a nightmare. Your astrologer knows it. He is eighty-seven, he has seen the dark and horrendous side of life. But finally, he says with Paul Brunton, 'There is peace behind the tumult, goodness behind the evil. Happiness behind the agony.'

Lastly, I am now revealing the secret of my predictions. We all know that we are in the Age of Aquarius. The Age of Aquarius will see the marriage of technology and humanity. The planets I personally take into active consideration are Jupiter, the planet of prosperity and plenty, and Saturn, the planet of duty as well as integrity. Both these planets will come together in the sign of Aquarius in 2081. Therefore intuition and common sense both dictate that it could well be the greatest year of the twenty-first century. The cynics are most welcome to say, 'Mr Daruwalla, you and many of us will not be there to verify the accuracy of this prediction.' My answer is, 'They are more than welcome to their own opinion. To each his own is the guiding principle of my life.'

The New Age Astrology

Gone are the days when you could blame it all on fate and karma. The New Age astrologers of the West would consider it a cop-out. New Age astrology is essentially about astrology of the Uranian age. Computers come into full play and cast your horoscope in a jiffy. They specialize in control and character analysis and assessment, shortcomings and strong points, and ways and means of improving yourself. It is all about a better *you*.

In a word, the New Age astrology makes you directly and almost solely responsible for your present state of affairs. It infiltrates the design of psychology, even psychiatry, meditation, yoga – in fact all the techniques of self-improvement and making you feel good about yourself.

Yes, I have not only attended the lectures at the Northern Virginia Astrological Association, the Network of Light, the Edgar Cayce Foundation, the Washington Astrology Forum, but have also spoken there by invitation. The refrain was: Feel good about yourself, no matter how poor your performance and how horrible your image, and how terribly miserable is your way of life.

It was all about pulling yourself up by your bootstraps. It was all about stress management. But above everything else, it was all about human relationships – and learning to let go, and move on, when a relationship had nothing more to offer, either because of death, decay or stagnation.

It was all about moving forward and taking charge of your life.

The New Age astrology is the astrology of the twenty-first century, the Uranian age, when the lion will lie with the lamb, when there will be faith and truth among people, when science will blend beautifully with religion, and when technology will mate with God! It is a bold, brave and beautiful vision of the future. The three outer planets, Uranus, Neptune and Pluto are the undisputed kings of the New Age astrology. Readers, have I whetted your appetite? If the answer is yes, I have succeeded! The New Age astrology is most certainly about stimulation and a consequent voyage of discovery on your own, with a friendly help from the astrologer.

Let me put it in another way. The New Age astrology is *not* about pandits and predictions, but about healing the body, mind and spirit, fluidity of relationships, personality development, and being the master of your own destiny.

Blessings and Special Wishes

Blessings are beautiful butterflies
Dancing right through the skies
Yes, God smiles
It seems that 2019 is the year of marriage of
many of my dear ones.

Notables are my darling Chirag, who is like a son to me, and his fiancée, Nikita. May all the energies bless them. Ganesha says they deserve the best.

Shiroy, the son of my great friend Behram Mehta will be tying the knot with Avasti who specializes in the world of fashion.

Raja Vikram Singh of Jambughoda was happy to discover that his daughter Chandra Mohini was pregnant after quite a long, long time. I had predicted that she would finally become pregnant. I must say that Raja Vikram Singh repaid me by treating me royally at his palace. He is a true king in every sense of the word. If you want to be as fit as a fiddle, I suggest you spend time at his place named, 'A home for Nature Lovers'. Fresh air, good home-made food, natural surroundings and a team of workers who will make you feel like being at a home away from home. I should know, I have been there at least five times.

My friend Jigar Shah will be celebrating the marriage of his daughter Lipi with Harsh in December 2018. My heartfelt blessings go with the entire family and specially with the newly-weds.

The very gracious couple Aspi Gandhi and Homai Gandhi celebrated the marriage of their daughter Pervin and her fiancé Jehan. It was a magnificent function. As I was hospitalized I could not attend the fine affair. Therefore Aspi and Homai graciously came to the hospital and took my blessings. The Gandhi family are my darlings.

People from Radio Mirchi came to my residence and held a large meeting of doctors to talk about health and hygiene in June 2018. I bless them for the good, great and grand work they are doing for humanity. They deserve a gold medal for sure.

Special Wishes for Dr Pallavi and Mr Pravin Darade

My dear noble daughter Pallavi,

May Lord Ganesha be with you and your family. I wish to analyse the successful woman achiever in you and I am very happy that a strong personality like you becomes a role model for success to all young daughters of our country. You are the perfect charming and intelligent beauty. Your positive attitude and strong determination defies the limitation of your astrological situations. However, it is my duty to inform you about predictions purely from astrological calculations.

Your Moon sign is Libra and ascendant is Gemini. Looking to the combinations of the planets I have come to the following calculations:

- Moon is weak as no planets are around it. This is known as Kemdrum yoga. Because of this you will earn very well. Investment in properties will be beneficial for you. Overall, your financial position will always be very good and there is nothing to worry about.

- After August 2018 and till December 2019 you will be successful in terms of money, but mentally you will be little unhappy.

- You will be in such professional work, that you will feel pressured.

- You should be careful in relationship issue and your child-related decisions during this time.

- Your health will be disturbed because of emotional and mental unhappiness.

- Rahu is in the eighth house and this is not good for your health. It gives gynaecological trouble, knee problems and arthritis too. Sometimes it gives complicated issues during such periods.

- You will have very good prestige in your office and in government position.

- Chances of promotions are very high after October 2018 and up to a period of one year.

- You will get fabulous growth in your life and career. Your life will be good, your fame and prestige will keep on growing.

- Every three to four years some issues or critical situations may come your way but it will be solved automatically. So there is nothing to worry about.

- You should be cautious about old known people and friends. Otherwise you will be in additional mental

BLESSINGS AND SPECIAL WISHES

and emotional problems till January 2020. Problems related to family disturbances will also be there during this period.

- After 2019, your time will be much better in the job environment. Your situation will improve. You will learn new things and will be able to handle extra load.

- From 2028 onwards you are likely to get involved in some kind of business. And this will greatly enhance your financial status.

- Financially you will be sound but you need to take some care of your health. Yoga, pranayam and some alternative healings will be helpful for you.

- You will have a long satisfied life. Old age will be good. Support from the child will be good. You will travel abroad too in old age.

- You should recite *Hanumanchalisa* and eleven times on Tuesdays and Saturdays.

- Wear the stone of Moon, that is, white pearl of 4 carats in a silver ring in the little finger.

- Fasting at least once on Wednesdays will be beneficial for you.

- For now you should start learning any new thing, this will be helpful to you.

- From the year 2021 you will start taking interest in spirituality.

- Overall, your life will be successful.

By the grace of Ganesha your horoscope shows scintillating success in your work area. In my language your family life will dance with joy, and what I have learnt or what life has taught me shows that mental happiness is the most important thing in life. So you will also be mentally very happy with your own input in your work, the work done by your noble husband to improve society at large, and also appreciation from the government.

The position of Sun in your chart shows that you will do great innovation, but at the same time you should take care of your back and legs. You should also do yoga and go for long walks and also go on vacation every year.

This is very important. As life progresses, your marital life will be happy, love will blossom between you, my daughter, and your husband, and believe me, there is nothing like having each other's company during old age. That will be your noble achievement.

In short, the coming six years really belong to you and I say *Tak dhina dhin* – that means triumph and glory by the grace of God.

Ganesha's Grace
Bejan Daruwalla
24 July 2018

My dear noble son Pravin,

May all the energies always be with you and your family. Your hard work is your strength. You are not born to be ordinary. Your profession and capabilities will give you extraordinary fame and prestige. Enjoy the sweet fruits of your passion but take extra care of your health and family. Your Moon sign is Aquarius and ascendant is

BLESSINGS AND SPECIAL WISHES

Scorpio. Saturn and Rahu are in the fifth house, causing *Pitrudosha*. That is *dosha* caused from the ancestors who don't get new birth and expect you to perform the prescribed rituals.

Because of that you had gone through lots of struggle and whatever you wish does not happen fast and takes a long time. The following are my findings and suggestions:

- You get everything after some hurdles.
- Whatever you wish and pursue with full efforts is delayed.
- You are moving ahead uncompromisingly and are meeting all your targets and schedules.
- You display awesome stamina and can move mountains during the coming period.
- In your life some pending issues are still there.
- To improve this you can do two things:
- Don't talk or discuss in details about your wish. You can give brief idea to the family members but never to any friend.
  - You many perform the puja of Narayan Bali at a place where three rivers meet. Usually, the places where people perform this puja are Chanod near Vadodara and Nashik. You may choose any of these two places. If this puja is performed, the ancestors get satisfied and bestow their blessings upon you. This is the most important remedy to nullify the dosha.

- o Regarding health, blood pressure, nerve-related issues and heart problems are the main things that should be taken care of.
- Besides taking care of your own health, you are also to take care of your wife, who is likely to have some health problem from April 2019 to the end of 2020.
- You have very good internal strength.
- You can work like anything if you decide.
- Once you determine, you will follow it up till you completely do it.
- But because of some planetary congestion like Saturn and Rahu in the fifth house, you may face some hurdles in your way, but you will be able to overcome it.
- Your growth will be very good.
- From January 2020, your financial growth will be very good.
- For remedy, you should wear the stone of Jupiter, that is, yellow sapphire weighing 5 carat on the index finger of your right hand in a gold ring on Thursday. It will be helpful for financial growth. Wear it from 2020 as Jupiter's period will start by then.
- Your social life will be good, but for the coming two years you need to take care of your health and also your relationship with your wife. Your wife's health will also be weak during that time.
- You should do work for spiritual progress because you have lots of potential for that. This will give you quiet peace, as can be seen from the position of your planets.

- Ganesha sees you turbo-charged as you successfully accomplish one task after another.

Last but not the least my best wishes are always with you.
    Ganesha's Grace

Bejan Daruwalla
30 July 2018

A Brush with Death

'The light that never was by land and sea,' says the Bible. Yes, I believe I may have had a glimpse of that light. How come? On 21 April, I had a brain stroke. It was sudden and had I not reached Avron Hospital right on the dot and been under the supervision and care of Dr Ranka I would have been in the next world.

The light I saw was pure white, but not dazzling and overpowering. Also it was translucent, that is, one could see through it. Now comes the strange part which I cannot explain. This light was within me. Also it was by my side. And lastly, the light could be seen for miles and miles, or so it seemed.

I openly admit that I did not see any light in the tunnel nor the bridge which people who are about to die see. Nor did I cross the stream. Most people claim to see their own bodies. Well, though I am an astrologer and a psychic, I did not see it at all. I can only tell you about what I actually saw.

There were fairies attending upon me. They said, 'He is so cute. But he shouts and acts like a child. It seems that men are hopeless at bearing pain and that men use their egos as an armour against the slings and arrows of fate.' They seemed to laugh at me but not mock me. Though I was strictly on the ventilator for two days, I feel I remonstrated against their remark that all men were children and couldn't bear pain. But I am not sure of it.

Here are a few suggestions which I am passing on after going through this ordeal or process or whatever else you would like to call it. The first and most important is not that I loved and wanted to unite with God. But that I now want everybody in life to start the race of life on an equal footing. Then it depends upon their individual potential and capacity. Secondly, I feel that in the final analysis great and universal compassion is the same as supreme consciousness. Lastly, I feel that the future of mankind is definitely into travelling in space and becoming at least half robots with artificial intelligence. In short it is my conviction that we will remain partly human and partly computers. I could say much more but I believe this more or less is my real finding after my very recent brush with death. I am neither sorry nor happy that I am alive and kicking. I believe that just to be what you are is enough. You may or may not agree with it because keeping an open mind is what pleases me the most.

About the Authors

Bejan Daruwalla

Ganesha says that though it may be true that Bejan Daruwalla is one of the 100 great astrologers in the last 1,000 years, he is perhaps the only one who has read the fortune of three prime ministers of India, namely, Morarji Desai, Atal Bihari Vajpayee and Narendra Modi. This Ganesha devotee at eighty-seven believes that goodness and compassion are much more important and certainly universal. He now believes in keeping a very low profile and tries to bless all people sincerely and with all the limited power that he has. In other words, in the wide, wide world, there is no substitute for sheer goodness. Goodness (including integrity) matters a million times more than greatness is Ganesha devotee Bejan's final word.

Nastur Daruwalla

How can I ever forget the scintillating prince, Lakshyaraj Singh of Mewar City Palace, Udaipur, who is like a son to me, as he has the divine blessings of Lord Ekling-ji of Shakti Peeth as the temple belongs to his great royal ancestors. If you have Lord Shiva with you what is more to say. My son Nastur stays in the heart of Prince Lakshyaraj Singh who is always ready to help the needy and downtrodden in life.

My son got blessings from the top industrialist of India, Shri Neeraj R. Bajaj, son of my patron Ramkrishna Bajaj for progress in life. I call Neeraj Bajaj my divine son as our relationship is at least fifty years old.

The coming back to power of the chief minister of Gujarat, Shri Vijay Rupani, was the bold and correct prediction of my son when all others were thinking that the chief minister will be changed. My son again got encouragement and blessings for progress in life and for writing this book, from the education minister of Gujarat, Shri Bhupendrasinh Chudasma, the home minister of Gujarat, Shri Pradipsinh Jadeja and the Union agriculture minister, Shri Parshottam Rupala.

My another beloved son and top industrialist of Mumbai, Shri Jimmey Mistry, gave his good wishes so that my legacy can continue in the form of my son Nastur.

Important Announcement

Our Ganesha devotee Bejan Daruwalla has moved from Mumbai to Ahmedabad. His Ahmedabad address is:

Bejan Daruwalla
Astrologer and Columnist
C/o Nastur Daruwalla
A-5, Spectrum Towers
Opposite Police Stadium
Shahibag
Ahmedabad 380004
India

Telephone
079-32954387
09825470377
08141234275

Email
info@bejandaruwalla.com
bejandaruwalla@rediffmail.com

Website: www.bejandaruwalla.com